Analytical NMR

Analytical NMR

Edited by

L. D. Field
Department of Organic Chemistry
University of Sydney
Australia

S. Sternhell
Department of Organic Chemistry
University of Sydney
Australia

JOHN WILEY & SONS
Chichester · New York · Brisbane · Toronto · Singapore

British Library Cataloguing in Publication Data available

ISBN 0 471 91714 1

Typeset by Associated Publishing Services, Ltd, Petersfield, Hants.
Printed in Great Britain by Bath Press, Bath, Avon.

List of Contributors

N. C. M. ALMA, Koninklijke/Shell Laboratorium Amsterdam, Badhuisweg 3, 1031 CM Amsterdam, The Netherlands

A. D. H. CLAGUE, Thornton Research Centre, Shell Research Ltd, PO Box 1, Chester CH1 3SE, UK

L. D. FIELD, Department of Organic Chemistry, University of Sydney, Sydney, 2006, N.S.W., Australia

P. W. KUCHEL, Department of Biochemistry, University of Sydney, Sydney, 2006, N.S.W., Australia

J. R. MOONEY, Standard Oil Company, Warrenville Center, Cleveland, OH 44128, USA

R.-D. REINHARDT, BASF AG, Central Research, Ludwigshafen, Federal Republic of Germany

C. E. SNAPE, Department of Pure and Applied Chemistry, University of Strathclyde, Glasgow, UK

M. SPRAUL, Bruker Analytische Messtechnik GmbH, Karlsruhe, Federal Republic of Germany

S. STERNHELL, Department of Organic Chemistry, University of Sydney, Sydney, 2006, N.S.W., Australia

Contents

CHAPTER 5: NMR of Zeolites, Silicates and Solid Catalysts
A. D. H. Clague and N. C. M. Alma

CHAPTER 6: Biological Applications of NMR
P. W. Kuchel

CHAPTER 7: Automatic NMR Analysis
M. Spraul and R.-D. Reinhardt

Analytical NMR
Edited by L. D. Field and S. Sternhell
© 1989 John Wiley & Sons Ltd

Chapter 1

Introduction

L. D. Field and S. Sternhell

Department of Organic Chemistry,
University of Sydney, Sydney, 2006, N.S.W., Australia

Although NMR spectroscopy can be considered to be an 'analytical' method in all of its applications, it was realised at an early stage[1-3] that in some areas, particularly those outside the mainstream structural and dynamic applications, the technique may be usefully categorized as 'Analytical NMR'.

As an analytical tool, NMR spectroscopy has a number of general characteristics. Its primary advantage is its versatility: the chromophores on which NMR spectroscopy depends are the nuclei of common isotopes and the nuclei of practically all elements of interest to chemists are amenable to study by NMR. Among the limitations of analytical NMR is the fact that some common elements (aluminium, oxygen, etc.) are difficult to observe owing to their nuclear properties. Further, NMR is inherently less sensitive than other analytical techniques although, with the increasing availability of superconducting magnets, advances in electronic components and routine Fourier transform (FT) instrumentation, lower sensitivity is becoming a less significant problem.

NMR spectroscopy is now an extremely broad subject and has evolved to encompass applications ranging from the simplest use for routine quality control to advanced, high-level pulsed experiments for the examination and analysis of complex mixtures. This book is an attempt to provide expert coverage of a number of areas in which NMR spectroscopy has found analytical applications. The choice of topics in this volume was governed both by the relative paucity of reviews in an area and by the availability of expert contributors. It is anticipated that this work will be extended in the future to cover other areas which fall into this category. The book is directed towards research workers who are familiar with the general principles and applications of NMR spectroscopy, but who would normally need to consult the primary literature to obtain an overview in these areas.

While a general familiarity with NMR spectroscopy was assumed by the Editors, it became clear that the amount of this assumed knowledge varied considerably between the various chapters in this collection. For this reason we have provided in Chapter 2 a brief review of the fundamentals of NMR, with particular emphasis on the principles and techniques actually referenced in subsequent chapters.

Chapter 3 is devoted to Quantitative Applications of ^{13}C NMR, a vastly undervalued method of general utility. Even though many of the principles of quantitative ^{13}C NMR had been discussed as early as 1974 by James Shoolery[4], very few laboratories have adopted ^{13}C NMR as an analytical method primarily because of the time which has been necessary to examine each sample. However, with the significant improvements in the sensitivity of modern NMR instrumentation, ^{13}C spectra are now routinely obtained in the range of minutes to a few hours and ^{13}C NMR is developing into a valuable analytical technique, particularly for the examination of complex carbon-based substances such as oils, coals and organic polymers.

Chapter 4 (Analysis of Fossil Fuels) examines the NMR methods which have been developed for the analysis of oils, oil shales and coals[5]. Chapter 5 (NMR of Zeolites, Silicates and Solid Catalysts) covers applications of multinuclear NMR to the characterization of the solid framework of common catalysts and molecules adsorbed and constrained within the catalyst structure. Both of these chapters describe the analysis of substances whose structures are difficult (or impossible) to assess by other means. Both chapters address areas of great and growing current technological interest.

Chapter 6 (Biological Applications of NMR) reviews applications of NMR to the study of actual biological systems (as distinct from the study of molecules of biological interest). This latter area of application encompasses all aspects of biology from the study of the mechanisms of fundamental cell processes to clinical evaluations and diagnosis.

Finally, Chapter 7 (Automatic NMR Analysis) gives an account, largely drawn from the authors' own work, of the development of methods for handling large numbers of samples. In an ever increasing range of applications, chemists involved with sample analysis demand NMR spectra as a routine piece of data on essentially all samples. Particularly in industrial environments but also in large graduate schools in universities, the problem of coping with the large number of samples is being addressed by automation involving the introduction of computer-controlled laboratory robots. The trend of all modern instrumentation towards the increasing use of computers for control, data storage and retrieval, data processing and sample analysis is an indication of the future in analytical NMR spectroscopy.

REFERENCES

1. F. Kasler, *Quantitative Analysis by NMR Spectroscopy*, Academic Press, New York, 1973.

2. D. E. Leyden and R. H. Cox, *Analytical Applications of NMR*, Wiley, Chichester, 1977.
3. M. L. Martin, J. J. Delpuech and G. J. Martin, *Practical NMR Spectroscopy*, Heyden, London, 1980.
4. J. N. Shoolery, *Varian Application Note*, NMR-73-4, Varian, Palo Alto, CA, 1974.
5. For recent reviews see, for example, (a) L. Petrakis and D. Allen, *NMR for Liquid Fossil Fuels*, Elsevier, Amsterdam, 1987; (b) M. A. Wilson, *NMR Techniques and Applications in Geochemistry and Soil Chemistry*, Pergamon Press, Oxford, 1987.

Analytical NMR
Edited by L. D. Field and S. Sternhell
© 1989 John Wiley & Sons Ltd

Chapter 2

Fundamental Aspects of NMR Spectroscopy

L. D. Field

Department of Organic Chemistry, University of Sydney, Sydney, 2006, N.S.W., Australia

1 INTRODUCTION

Nuclear magnetic resonance (NMR) spectroscopy[1] can be defined as the absorption and emission of electromagnetic radiation by the nuclei of certain atoms when they are placed in a magnetic field. In order to absorb electromagnetic radiation, nuclei must possess a non-zero magnetic moment. This is the case for any nucleus with a non-zero nuclear spin quantum number ($I \neq 0$) and, although the majority of the elements in the Periodic Table have at least one isotope capable of nuclear magnetic resonance, there are a number of abundant isotopes (e.g. ^{12}C, ^{16}O, ^{32}S) which have $I = 0$ and are not detectable by NMR.

Samples for NMR spectroscopy are typically liquids (or solutions) and solids. NMR has been applied only rarely to gaseous samples. For NMR spectroscopy,

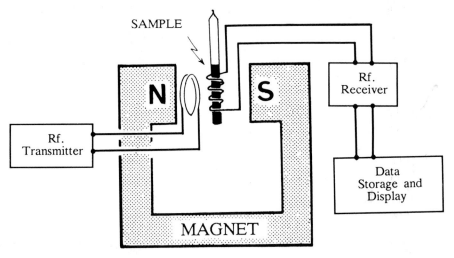

Fig. 1. Schematic diagram indicating the basic components of an NMR spectrometer

the sample must be placed in a magnetic field before the absorption of electromagnetic radiation can occur. The basic components of any NMR spectrometer include a strong magnet, into which the sample is placed, a radiofrequency transmitter and a receiver system connected to some type of data display or storage device (Figure 1).

The ^{1}H nucleus (i.e. the proton) is by far the most commonly studied nucleus by NMR spectroscopy, primarily because of the ease of observation, its high natural abundance and the fact that it is invariably present in the majority of samples. Despite its low natural abundance (1.1%), ^{13}C is also an important nucleus because carbon forms the backbone of all organic compounds and structural information can be obtained by NMR spectroscopy. ^{19}F and ^{31}P are frequently examined for the specific classes compounds which contain these elements. With modern instrumentation, NMR spectra can be obtained routinely on most isotopes where NMR spectroscopy is possible[2].

2 THE NMR SPECTRUM

An NMR spectrum is normally presented as a graph (or table) of absorption intensity against the frequency of radiation absorbed by the nuclei in a sample. NMR spectroscopy is quantitative in that the integrated intensity of a signal is proportional to the concentration of nuclei giving rise to it and for this reason NMR spectroscopy is a powerful technique for establishing the relative concentrations of components in mixtures.

The frequency (or frequencies) of radiation absorbed by nuclei in an NMR experiment fall within the radiofrequency region of the electromagnetic spectrum and depend on a number of factors, including (i) the strength of interaction of the nuclei with the magnetic field in which the sample is placed (the Zeeman interaction); (ii) the intramolecular interaction of the nucleus with other magnetic nuclei in the molecule via the bonds of the molecule (scalar or J coupling); (iii) the 'through-space' intramolecular or intermolecular interaction between magnetic nuclei (dipolar or D coupling); and (iv) the influence of a nuclear quadrupole moment for nuclei with $I > \frac{1}{2}$ (quadrupole coupling).

The Zeeman interaction and scalar, dipolar and quadrupolar coupling can be treated in depth[3], but a detailed description is unnecessary in order to be able to understand the fundamental concepts behind most analytical applications of NMR. Of the factors listed above, the Zeeman interaction is the most important, because it is several orders of magnitude larger than the others. Dipolar and quadrupolar coupling are important factors which influence the NMR spectra of solids and oriented media (e.g. liquid crystalline solutions[4]) but they do not affect the spectra of mobile liquids or molecules dissolved in non-viscous solutions.

2.1   The Zeeman interaction and the Larmor equation

When placed in a magnetic field, a nucleus of spin I can adopt any of $2I + 1$ states and the interaction of the nucleus with electromagnetic radiation causes transitions between the states (Figure 2). The frequency (v_0) of electromagnetic radiation absorbed by nuclei depends primarily on the strength of the magnetic field ($\mathbf{B_0}$) in which the sample is placed and on the magnetogyric ratio (γ) of the nuclei. The quantities γ, v_0 and $\mathbf{B_0}$ are related by the Larmor equation:

$$v_0 = \frac{\gamma}{2\pi} \mathbf{B_0} \tag{1}$$

Magnets with field strengths up to 14.1 tesla (T) are currently available commercially for NMR spectroscopy. This strength corresponds to absorption frequencies up to 600 MHz (i.e. in the radiofrequency region of the electromagnetic spectrum). The resonance frequencies for a number of commonly observed nuclei are given in Table 1. Note that, in a given magnetic field, there is a significant difference between the resonance frequencies of different types of nuclei (typically tens to hundreds of MHz) and signals for any specific isotope can be observed by choosing the appropriate frequency range.

The energy difference (ΔE) between the states which a nucleus can adopt in magnetic fields which are accessible to NMR spectroscopy is of the order of 10^{-27}–10^{-25} J and the difference in population between the various states (as described by the Boltzmann equation) is very small—typically about 1 part in 10^5. Consequently, NMR signals are weak in comparison with those observed with other forms of spectroscopy which have higher ΔE. From the Larmor equation it can be seen that ΔE (and hence the population difference and intensity of the potential NMR signal) increases with increasing magnetic field

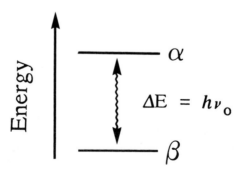

Fig. 2. The two energy states available to a nucleus with $I = \frac{1}{2}$. The absorption of r.f. radiation at the correct frequency (v_0) induces transitions between the states

TABLE 1
NMR Properties for a number of commonly observed nuclei

Isotope	Spin	Natural abundance (%)	NMR frequency[a] (MHz)	NMR frequency[b] (MHz)	Usual reference compound
^1H	$\frac{1}{2}$	99.985	100.00	500.00	TMS
^{10}B	3	19.58	10.75	53.73	$BF_3.OEt_2$
^{11}B	$\frac{3}{2}$	80.42	32.08	160.42	$BF_3.OEt_2$
^{13}C	$\frac{1}{2}$	1.108	25.14	125.72	TMS
^{14}N	1	99.63	7.22	36.12	NH_3, NO_3^- or $MeNO_2$
^{15}N	$\frac{1}{2}$	0.37	10.13	50.66	NH_3, NO_3^- or $MeNO_2$
^{17}O	$\frac{5}{2}$	0.0037	13.56	67.78	H_2O
^{19}F	$\frac{1}{2}$	100.0	94.08	470.39	CCl_3F
^{23}Na	$\frac{3}{2}$	100.0	26.45	132.26	1 M $NaCl-H_2O$
^{27}Al	$\frac{5}{2}$	100.0	26.06	130.29	$[Al(H_2O)_6]^{3+}$
^{29}Si	$\frac{1}{2}$	4.7	19.86	99.32	TMS
^{31}P	$\frac{1}{2}$	100.0	40.48	202.40	85% H_3PO_4
^{35}Cl	$\frac{3}{2}$	75.53	9.80	48.99	$NaCl-H_2O$
^{59}Co	$\frac{7}{2}$	100.0	23.61	118.07	$[Co(CN)_6]^{3-}$
^{119}Sn	$\frac{1}{2}$	8.58	37.27	186.36	Me_4Sn
^{195}Pt	$\frac{1}{2}$	33.8	21.50	107.49	$PtCl_6^{2-}$ or $[Pt(CN)_6]^{2-}$
^{207}Pb	$\frac{1}{2}$	22.6	20.92	104.61	Me_4Pb

[a] In a magnetic field of 2.3488 T.
[b] In a magnetic field of 11.744 T.

strength. This has been one factor fuelling the development of NMR spectrometers with magnets with higher field strengths (and hence operating at higher
frequencies).

2.1.1 Shielding and chemical shift

The Larmor equation provides a coarse discrimination between the nuclei of
different isotopes; however, the principal applications of NMR spectroscopy
examine differences between nuclei of the *same isotope* but which occupy
different chemical environments. The ability of NMR spectroscopy to distinguish between nuclei of the same type arises because nuclei placed in a magnetic
field are *shielded* or *screened* from the field by the electrons which surround
them. The magnetic field actually experienced by a nucleus (i.e. the 'effective
field' **B**) is the applied field $\mathbf{B_0}$ modified by the screening effect of electrons. The
degree of screening depends explicitly on the electron density and therefore on
the nature of the bonding in the molecule in which the nuclei are embedded. The
different screening experienced by nuclei in different chemical environments is
termed the *chemical shift*. Unless they are in chemically identical environments
(i.e. related by some element of symmetry), different nuclei are shielded to

TABLE 2
Typical chemical shift ranges for protons and carbon
nuclei in organic compounds (chemical shifts in ppm
from TMS)

Proton environment	^{1}H chemical shift (ppm)	Carbon environment	^{13}C chemical shift (ppm)
Aliphatic —C—H	0.2–1.5	Aliphatic C	−2.5–35
Terminal =C—H	4.5–8.0	Alkenes =C	110–150
Aromatic ϕ—H	6.5–8.0	Aromatic C	120–150
Terminal C≡C—H	2.4–3.0	Alkyne ≡C	65–90
Aldehydic O=C—H	9.5–10.0	Carbonyl C=O	160–220
CH$_3$—O—	3.4–4.0	CH$_3$—O—	50–60

differing extents and hence have different resonance frequencies. The resonance
frequency (v) of a nucleus can be calculated by substituting the effective field **B** in
to the Larmor equation:

$$v = \frac{\gamma}{2\pi} \mathbf{B} \qquad (2)$$

where $\mathbf{B} = \mathbf{B_0}(1 - \sigma)$ and σ is a dimensionless fraction termed the *shielding
constant*. Note that equation 2 indicates that the *more shielded* a nucleus from
the applied magnetic field, the *lower* is its resonance frequency.

The resonance frequency of a nucleus (expressed in Hz) as determined by
equation 2 is directly proportional to $\mathbf{B_0}$ and changes from spectrometer to
spectrometer, depending on the strength of the magnet. It is more convenient to
express v in dimensionless units parts per million (ppm) by dividing by v_0
($\approx \gamma \mathbf{B_0}/2\pi$). NMR frequencies expressed in ppm are given the symbol δ, they are
independent of $\mathbf{B_0}$ and are tabulated as characteristic molecular properties.
Typical chemical shifts for protons and carbon nuclei in common organic
compounds are listed in Table 2 and an example of a simple spectrum is given in
Figure 3.

Resonance frequencies (or shieldings) are typically measured relative to the
frequency of some standard compound, taken by convention as a reference.
Chemical shifts are usually expressed in units of ppm from the resonance of the
reference compound. Reference compounds for various commonly observed
isotopes are included in Table 1.

2.1.2 Chemical shifts in solids

The shielding of a nucleus depends on the orientation of the molecule and its
bonds with respect to $\mathbf{B_0}$. In the spectra of liquids and isotropic solutions,
molecular reorientation is sufficiently rapid that shielding is averaged over all

^{1}H NMR Spectrum

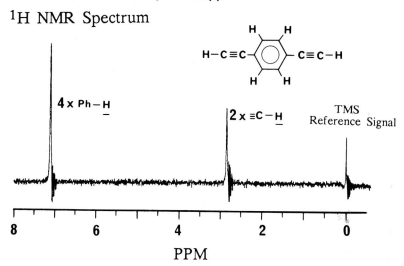

Fig. 3. ^{1}H NMR spectrum of 1,4-diethynylbenzene recorded in a magnet of 1.409 T (the Larmor frequency for protons is 60 MHz). There are two resonance signals (corresponding to the aromatic and acetylenic protons) and their integrated intensities are in the ratio 4:2

orientations. However, in the spectra of solids where the molecules are not free to rotate, the chemical shift of a nucleus depends on the orientation of the molecule in the magnetic field. The directional dependence of nuclear shielding is termed the chemical shift anisotropy (CSA). The chemical shifts in the NMR spectrum of a single crystal of a solid vary with the orientation of the crystal; the spectra of powders and non-crystalline solids are usually broad envelopes which result from the overlap of spectra from molecules with all possible orientations with respect to the magnetic field.

The shielding constant, σ, is not scalar quantity but rather a second-rank tensor. For a nucleus in some molecule *fixed* in a magnetic field (e.g. in a stationary solid), one can express the shielding in any direction in terms of a shielding tensor:

$$\sigma = \begin{vmatrix} \sigma_{xx} & \sigma_{xy} & \sigma_{xz} \\ \sigma_{yx} & \sigma_{yy} & \sigma_{yz} \\ \sigma_{zx} & \sigma_{zy} & \sigma_{zz} \end{vmatrix} \tag{3}$$

By choosing a suitable coordinate, system, the shielding tensor can be diagonalized to a matrix with only three non-zero components (with principal elements σ_{xx}, σ_{yy}, σ_{zz}). The anisotropy of the chemical shift of a nucleus (i.e. the difference between principal elements of the shielding tensor) can often be of the order of several kilohertz and, for nuclei in single crystals of pure compounds, the components of the shielding tensor can be measured from the orientation dependence of the chemical shifts.

For molecules in mobile solution, the averaged nuclear shielding observed is the trace of the shielding tensor, i.e. $\sigma_{isotropic} = \frac{1}{3}(\sigma_{xx} + \sigma_{yy} + \sigma_{zz})$.

2.1.3 Magic angle spinning

The orientation dependence of the chemical shifts of nuclei in solids can be suppressed by sample spinning. It has been shown[5,6] that if a solid is spun rapidly, the observed shielding for any nucleus in the sample can be expressed by

$$\sigma_{obs} = \tfrac{1}{2} \sin^2 \theta (\sigma_{xx} + \sigma_{yy} + \sigma_{zz}) + \tfrac{1}{2} A(3 \cos^2 \theta - 1) \qquad (4)$$

where A is a term which depends on the chemical shift tensor and the orientation of its principal axes with respect to the spinning axis and θ is the angle of the spinning axis with respect to $\mathbf{B_0}$. When θ is set to 54.7° (the *magic angle*), equation 4 reduces to $\sigma_{obs} = \frac{1}{3}(\sigma_{xx} + \sigma_{yy} + \sigma_{zz})$, i.e. to the isotropic value, irrespective of the orientation of molecules with respect to the spinning axis. Consequently, the spectra of solid samples acquired with magic angle spinning (MAS) do not exhibit the broadness arising from the overlap of spectra arising from nuclei in molecules with different orientations.

To remove totally the effect of CSA from a spectrum, the rate of magic angle spinning must be large compared with the chemical shift anisotropy (typical rotation speeds are 1–5 kHz). Since the chemical shift increases with increasing $\mathbf{B_0}$, higher rotation speeds are necessary with higher magnetic fields. If the spinning rate is not sufficiently rapid to average CSA completely, spinning sidebands result.

2.2 Scalar coupling

Scalar coupling (also termed spin–spin coupling, J coupling or indirect coupling) is a consequence of the magnetic interaction between nuclei with spin (i.e. those with $I \neq 0$) transmitted *via the bonding electrons* of a molecule. The basic resonance frequency of a nucleus can be perturbed by the interaction with other magnetic nuclei, giving rise to multiplet splittings in NMR spectra.

2.2.1 Signal multiplicity

The NMR signal of a nucleus coupled to **n** equivalent nuclei with spin I will be split into a multiplet with $(2\mathbf{n}I + 1)$ lines. If $I = \frac{1}{2}$, the relative intensity of the lines in the multiplet will be given by the binomial coefficients of order $n - 1$:

n	Multiplicity		Relative line intensities
0	1	singlet	1
1	2	doublet	1 : 1
2	3	triplet	1 : 2 : 1
3	4	quartet	1 : 3 : 3 : 1
4	5	quintet	1 : 4 : 6 : 4 : 1
⋮	⋮		⋮ ⋮ ⋮ ⋮ ⋮
etc.	etc.		etc.

A ^{13}C nucleus coupled to three protons (i.e. as in a CH_3 group) will appear as a quartet in the ^{13}C NMR spectrum. Similarly, the ^{13}C signal of a methylene group will appear as a triplet and the signal of a methine group as a doublet. If a nucleus is coupled to *one* nucleus with $I > \frac{1}{2}$, the multiplet observed always has lines with equal intensity. If a nucleus is coupled to n equivalent nuclei with $I > \frac{1}{2}$, the intensity distribution of the lines in the multiplet is more complex.

2.2.2 The magnitude of the coupling constant

The coupling constant J (usually expressed in Hz) is the measure of the capacity of the nucleus to sense the spin state of another nucleus *through the bonds* of the molecule. Spin–spin coupling is a molecular property since it depends on the nature of the bonds separating the coupled nuclei. The magnitude of spin–spin coupling is *independent of the magnetic field strength* of the NMR spectrometer in which the spectrum is recorded.

The magnitude of coupling constants generally decreases as the number of bonds between the coupled nuclei increases (Table 3). By convention, a superscript before the symbol J represents the number of intervening bonds between the coupled nuclei. The magnitude of the coupling depends explicitly on the nature of the intervening bonds and unexpectedly large couplings exist

TABLE 3
Typical coupling constants (Hz) for commonly en-countered pairs of nuclei

Compound	Coupling constant (Hz)	Compound	Coupling constant (Hz)
CH_3CH_3	$^1J_{CH} = 125$	CH_3F	$^1J_{CF} = 157$
$CH_2{=}CH_2$	$^1J_{CH} = 156$	CH_3CH_3	$^1J_{CC} = 35$
$CH{\equiv}CH$	$^1J_{CH} = 249$	$(CH_3)_2PH$	$^1J_{PH} = 207$
$CH{\equiv}CH$	$^1J_{CC} = 172$	$(CH_3CH_2)_3P$	$^4J_{PH} = 0.5$
C_6H_6	$^1J_{CH} = 159$	$(CH_3)_3P$	$^3J_{PH} = 2.7$
$CH_3CH_2CH_3$	$^3J_{HH} = 7.2$	PF_3	$^1J_{PF} = 1410$

across many bonds if there is a particularly favourable bonding pathway (e.g. extended π conjugation or a favourable rigid σ-bonded skeleton).

The coupling constant can have either a positive or a negative sign. The sign of the coupling constant generally has no visible effect on the appearance of NMR spectra (at least in first-order spin systems). The coupling between geminal protons is characteristically negative and the coupling between vicinal protons typically has a positive sign.

In similar bonding environments, the coupling is proportional to the product of the magnetogryic ratios of the coupled nuclei, i.e. $J_{ij} \propto \gamma_i \gamma_j$. This relationship is important in considering the NMR spectra of molecules containing different isotopes of the same element, e.g. coupling between a nucleus and ^{15}N is approximately 1.4 times as large as the coupling to ^{14}N (since $\gamma_{15N}/\gamma_{14N} \approx 1.4$).

2.2.3   Spin decoupling

Spin decoupling results from the application of an r.f. field exactly at the resonance frequency of one nucleus while observing the NMR spectrum of nuclei which are coupled to it. The r.f. irradiation must be sufficiently strong to cause rapid flipping of the irradiated nucleus, so that nuclei coupled to it cannot sense its spin state and any splitting due to coupling to the irradiated nucleus disappears from the spectrum. The irradiation power (B_2) necessary to completely decouple one nucleus from another depends on the size of the coupling constant and the magnetogyric ratio of the irradiated nucleus: $B_2 > 2\pi J/\gamma$.

Selective decoupling.

Provided one can control the power of the irradiating r.f. field sufficiently precisely, it is possible to decouple one particular nucleus frequency without significantly affecting others at nearby frequencies. The term 'homonuclear decoupling' is used to describe a selective decoupling experiment where the type of nucleus decoupled is the same as that being observed (e.g. decoupling one proton while observing other protons). The term 'heteronuclear decoupling' refers to the situation where the type of nucleus irradiated is different to that being observed (e.g. decouple ^1H while observing ^{13}C).

Selective decoupling is particularly important in analyses where the measurement of coupling constants is necessary (e.g. in determining the structure of an unknown organic compound). Selective decoupling can aid in unravelling the complex multiplicity of a signal by selectively removing splitting from one coupled nucleus at a time.

Broadband or noise decoupling.

With heteronuclear decoupling it is possible and usually desirable to decouple completely all nuclei of one type while observing the spectrum of another (e.g. decouple all ^{1}H nuclei while observing the ^{13}C spectrum). Broadband decoupling is usually achieved by noise or square-wave modulation of the decoupling r.f. irradiation, effectively resulting in a spread of frequencies irradiated simultaneously. In more modern NMR spectrometers, broadband decoupling is achieved by pulsed methods (see Section 6.1.5).

Broadband decoupling removes *all* splittings which would normally result from coupling to one nuclear species. In a ^{13}C NMR spectrum, in the absence of ^{1}H decoupling, CH_3 groups appear as quartets, CH_2 groups appear as triplets, etc., owing to coupling of the carbon nuclei to the directly bonded protons. With broadband ^{1}H decoupling all splittings due to ^{1}H coupling are removed so all signals appear as singlets. For isotopically rare nuclei (such as ^{13}C) where the NMR signal is inherently weak, decoupling to collapse multiplets to a singlet or signal of lower multiplicity results in a dramatic increase in signal height and hence the ability to detect signals above the background noise. In any heteronuclear situation where coupling information is not required, spectra are invariably acquired with broadband ^{1}H decoupling.

Off-resonance decoupling.

By application of an r.f. field not exactly at the Larmor frequency of a nucleus it is possible to 'partially' decouple it from other nuclei. The principal application of off-resonance decoupling is in determining the multiplicity of signals in the spectra of heteronuclei with coupling to ^{1}H. Off-resonance irradiation of the ^{1}H spectrum results in a heteronuclear spectrum in which the splittings are reduced from the full value of the coupling constants but which retain the correct multiplicities (quartets, triplets, etc.) and can hence be assigned as CH_3, CH_2, etc.

2.3   Dipolar coupling

Dipolar or direct coupling results from the dipole–dipole (through space) interaction between nuclei with spin. For two nuclei i and j, the dipolar coupling constant D_{ij} is a measure of the strength of the dipolar interaction between them and depends on their internuclear distance (r_{ij}), their magnetogyric ratios (γ_i and γ_j) and the orientation (θ_{ij}) of the internuclear vector with respect to the applied magnetic field:

$$D_{ij} = \frac{h\gamma_i\gamma_j}{8\pi^2}\left(\frac{3\cos^2\theta_{ij} - 1}{r_{ij}^3}\right) \tag{5}$$

The magnitude of dipolar coupling is much greater than scalar coupling and both intermolecular and intramolecular dipolar coupling are important. Dipolar coupling is a dominant factor in determining the NMR spectrum of solids and anisotropic liquids; however, rapid molecular reorientation averages θ such that dipolar coupling is zero, and in mobile liquids and molecules in isotropic solution, dipolar coupling makes no contribution to the NMR spectrum.

D_{ij} decreases dramatically as r_{ij} increases. D_{ij} increases in proportion to the magnetogyric ratio of each of the coupled nuclei so that coupling is strongest between high-γ nuclei (e.g. between two ^1H nuclei). In principle, dipolar coupling can be removed by magic angle spinning (D_{ij} is zero if the average value of $\langle 3\cos^2\theta_{ij} - 1\rangle$ is zero). However, in practice the required spinning speeds are too great (typically 50–70 kHz for H–H and C–H dipolar coupling) for MAS to be employed routinely. Heteronuclear dipolar coupling (e.g. between C and H) can be removed by high-power decoupling of one species (e.g. observe ^{13}C while decoupling ^1H). Since dipolar couplings are much larger than scalar interactions, the power required for decoupling is correspondingly greater. H–H dipolar interactions can be removed or suppressed by the use of pulsed NMR techniques (see Section 6.1.4).

2.4   Quadrupolar effects

For nuclei with spin greater than $\frac{1}{2}$ (e.g. ^{17}O, ^{27}Al, ^{23}Na, ^{11}B), the interaction of the magnetic moment of the nucleus with an unsymmetrical electric field gradient gives rise to additional splittings in the NMR spectra. As with dipolar coupling, the magnitude of quadrupolar splittings is dependent on orientation and consequently quadrupolar splitting is not observed in the spectra of liquids or molecules in mobile solution.

Quadrupolar nuclei in solids and anisotropic liquids may exhibit splittings or line broadening due to quadrupolar coupling.

3 T_1 RELAXATION

When placed in a magnetic field, the nuclear spins are aligned by $\mathbf{B_0}$ and the spin population is distributed amongst the various states available according to a Boltzmann distribution. At equilibrium there are more spins aligned in the direction of $\mathbf{B_0}$ than against it, with the consequence that the sample is slightly magnetized in the direction of the magnetic field. Magnetization in the direction of the applied magnetic field is termed *longitudinal magnetization*. The nuclear spins are not stationary in the magnetic field, but precess about the direction of $\mathbf{B_0}$ with a characteristic frequency, $\omega = -\gamma \mathbf{B_0}$.

Absorption of r.f. radiation at or near the Larmor frequency causes a redistribution of the spins amongst the states and this gives rise to a non-equilibrium situation. The population recovers to an equilibrium distribution by a first-order (exponential) process with a time constant defined as T_1 (the longitudinal relaxation time):

$$\Delta p = \Delta p_{eq} \exp\left(-T_1/t\right) \qquad (6)$$

where Δp is the population difference between two states following a disturbance and Δp_{eq} is the equilibrium population difference. Note that T_1 is a measure of the time it takes for the *populations* of the spin states to recover to their equilibrium values and a measure of the time it takes for the sample to recover its equilibrium magnetization in the (longitudinal) direction of $\mathbf{B_0}$. A number of other relaxation times can be defined for a spin system (e.g. T_2, $T_{1\rho}$), and these refer to the rates of relaxation of other quantities in the spin system (see Section 5.2.2).

Depending on the mechanisms available to relax the nuclei, T_1 can have values ranging from microseconds (efficient relaxation) to several hours (inefficient relaxation). The relaxation time of nuclei in the sample has a number of important consequences:

(i) The observable signal in an NMR experiment depends on the population difference between the states available to the nuclei. The maximum signal is observed when the nuclei in the sample are fully relaxed (i.e. the spins are at equilibrium and the population difference is that given by the Boltzmann equation).

  The relaxation time is a factor which determines how often the NMR experiment can be repeated. If a number of NMR measurements are made on a sample with an insufficient delay between spectral acquisitions, nuclei in the sample will not be fully relaxed following one experiment before the beginning of the next and consequently the intensity of the observable signal will decrease with successive acquisitions. This is particularly important for pulsed NMR experiments (see Section 5.2), where it is possible to record an NMR spectrum in a fraction of a second and desirable to record a number of spectra and add them to improve the signal.

  If a sample is irradiated continuously with r.f. radiation at the Larmor frequency, the populations of the states rapidly equalize. In this situation, no NMR signal is observable ($\Delta p = 0$) and the nuclei are said to be 'saturated'. Saturation can be a useful technique for the observation of weak NMR signals (from species present in relatively low concentration) in the presence of overwhelming strong signals. Saturation of the nuclei giving the strong signal, removes its signal from the spectrum and permits the observation of other species.

(ii) T_1 contributes to the linewidth of each NMR signal. If the nuclei return to equilibrium rapidly (i.e. rapid relaxation) it is not possible to measure the resonance frequency accurately. This can be viewed as a consequence of the

Heisenberg uncertainty principle : to measure an energy difference (and hence the nuclear resonance frequency in NMR spectroscopy) accurately, one needs a long time. If the system relaxes rapidly, the time available is small and hence ΔE ($= h\nu$) is poorly defined and the observed NMR resonances are broad.

The linewidth of a signal is inversely related to T_1 the linewidth of a resonance is inversely proportional to T_2, which by definition must be less than (or equal to) T_1. If a nucleus relaxes quickly (T_1 is small) then the NMR signals for that nucleus are broad (i.e. the linewidth is large) and poorly defined. If a nucleus relaxes slowly (T_1 is large) then the NMR signals for that nucleus are sharp and well defined.

(iii) T_1 values are particularly sensitive to molecular motion and can be used to derive information about the mobility and dynamics of whole molecules or segments of molecules.

(iv) The measured value of T_1 is a valuable structural parameter, characteristic of the type of nucleus and its environment. T_1 values can be tabulated and correlated with chemical and structural properties in the same way as chemical shifts and coupling constants.

3.1   Important mechanisms for T_1 relaxation

T_1 values depend on the mechanisms which are available to a nucleus to induce transitions and help regain equilibrium. T_1 depends on many factors, including the state of the sample (solid, liquid or gas), molecular size and mobility, temperature, solvent and the presence and nature of impurities. The most important factors which contribute to T_1 relaxation include the following.

3.1.1   Dipole–dipole interactions

Dipole–dipole relaxation is the result of interaction of nuclei with other magnetic dipoles in the sample. The dipolar contribution to T_1 (commonly given the symbol T_{1DD}) arises because the tumbling motion of magnetic nuclei gives rise to fluctuating magnetic and electric fields which can stimulate transitions and assist the spins relax to regain their equilibrium population distributions.

T_{1DD} is inversely proportional to the time for molecular reorientation (τ_c):

$$T_{1DD} \propto 1/\tau_c \qquad (7)$$

The dependence of T_{1DD} on τ_c is a consequence of the fact that the efficiency with which dipoles interact (and hence relax each other) depends on their relative orientation. If dipoles reorient rapidly with respect to each other, they have little opportunity to interact, and conversely, if the reorientation is slow, dipole–dipole relaxation processes are efficient.

Small molecules (MW < 1000) in solution tumble rapidly and mobility is not a primary determinant of T_1. However, for large molecules (polymers, and biopolymers such as proteins and nucleic acids) where molecular tumbling is slow τ_c is long and the nuclei in these molecules relax rapidly (and consequently give very broad signals in their NMR spectra). The nuclei of molecules in viscous solutions and solids give very broad NMR spectra due to efficient dipole–dipole relaxation.

In molecules consisting of a rigid backbone with attached side-chains, nuclei attached to the side-chains have considerably more freedom and mobility than those on the backbone. Typically, nuclei on the molecular backbone (e.g. on the peptide backbone of proteins or polymers) have shorter T_1 values (i.e. relax more efficiently) than those which are part of mobile side-chains.

T_{1DD} is inversely proportional to the sixth power of the distance between the dipoles (r) (i.e. $T_{1DD} \propto 1/r^6$), and therefore nuclei which are close together in space, relax each other most efficiently. For nuclei which are 'buried' inside a molecule (e.g. ^{13}C), relaxation is governed primarily by those nuclei which are directly bound, i.e. as close as possible. ^{13}C nuclei with no directly bound protons (quarternary carbons) typically have comparatively longer T_1 values than protonated carbon nuclei.

3.1.2   Paramagnetic relaxation

For nuclei which are paramagnetic (i.e. possess an unpaired electron) the magnetic moment of the electron provides an efficient means of relaxation for the nucleus. In any sample, the presence of paramagnetic species, including oxygen and many transition metal ions (Fe^{3+}, Cr^{3+}, Mn^{2+}, etc.), is particularly important because such species have a large magnetic moment and cause rapid (efficient) relaxation of nuclei. In some instances, the salts of paramagnetic metals are added to a sample to enhance the rate of relaxation of nuclei with inconveniently long T_1 values. On the other hand, in high-resolution NMR spectroscopy, where the aim is to achieve the narrowest possible resonance, it is imperative that all paramagnetic species (including oxygen) are excluded from the sample.

3.1.3   Quadrupolar relaxation

Nuclei with $I > \frac{1}{2}$ possess a nuclear quadrupole which can relax the nucleus extremely efficiently. The NMR signals from quadrupolar nuclei are often several kilohertz wide and for some quadrupolar nuclei the NMR signals are so broad that spectra are unobservable.

3.1.4   Other contributions to T_1 relaxation

There are numerous other mechanisms which can contribute to T_1 relaxation for nuclei in specific classes of molecules or groups. For small molecules or groups of high symmetry, rapid rotation can contribute to T_1 relaxation (termed a *spin rotation* contribution to T_1). For nuclei which are part of (or close to) highly anisotropic groups (e.g. ^{13}C in carbonyl groups or multiple bonds) chemical shift anisotropy (see Section 2.1.2) can also provide a fluctuating magnetic field to stimulate relaxation if molecular motion is of an appropriate time scale.

4   THE NUCLEAR OVERHAUSER EFFECT (NOE)

The nuclear Overhauser effect (NOE) results in a change in *intensity* (increase or decrease) of the NMR signal of a nucleus when the resonance of some other nucleus in the molecule is *saturated*. The most important general consequence of the NOE is that the intensity of NMR signals can be enhanced significantly by saturation of some of the nuclei in a sample. (For a more detailed description of the NOE and its applications, see Refs 7–9.)

4.1   Signal enhancement by NOE

The NOE arises via dipole–dipole interactions (i.e. through space) between the saturated and observed nuclei. If the relaxation of a nucleus a occurs *entirely* via dipolar interaction with one other b, the maximum enhancement of the signal of a on saturation of b depends on the relative magnetogyric ratios of the saturated and observed nuclei according to the equation

$$\text{Maximum NOE enhancement factor} = \frac{\gamma_{sat}}{2\gamma_{obs}} \qquad (8)$$

where γ_{obs} is the magnetogyric ratio of the nucleus observed and γ_{sat} is the magnetogyric ratio of the nucleus saturated. In any homonuclear experiment (i.e. the same type of nucleus is observed as is saturated), the maximum possible NOE enhancement factor is 0.5 (i.e. the intensity of the observed signal is increased by 50%). However, significant increases in the intensity of signals from low magnetogyric ratio nuclei can be obtained if they receive an NOE from a nucleus with a higher magnetogyric ratio (Table 4). The intensity of ^{13}C resonances is increased by up to 200% if they receive an NOE enhancement from saturated protons. Note that for some important nuclei (e.g. ^{15}N, ^{29}Si) the

TABLE 4
Maximum NOE enhancements for various combinations of observed and saturated nuclei

Observed	Saturated			
	^{1}H	^{19}F	^{31}P	^{2}H
^{1}H	0.50	0.47	0.20	0.077
^{19}F	0.53	0.50	0.21	0.082
^{31}P	1.24	1.16	0.50	0.19
^{29}Si	−2.52	−2.37	−1.02	−0.39
^{13}C	1.99	1.87	0.80	0.30
^{15}N	−4.93	−4.64	−2.00	−0.76

NOE is negative, i.e. ^{15}N signals are decreased by the saturation of protons. However, the magnitude of the observed ^{15}N signal is increased by NOE (even though the signal changes sign) and the spectra have a significantly improved absolute intensity.

The NOE differs from nucleus to nucleus in a molecule—all nuclei do not receive the same enhancement. In a heteronuclear experiment, saturation of the ^{1}H spectrum causes most enhancement to the protonated atoms, i.e. those to which protons are directly bound. Consequently, integration and quantification of heteronuclear spectra are not reliable under conditions where the NOE operates. This is particularly important where decoupling is employed to simplify heteronuclear spectra (see Section 2.2.3) since, if r.f. irradiation for decoupling operates for a period greater than a few T_1, the decoupled nucleus will be saturated and provide an NOE to nearby nuclei.

Gated decoupling is a technique commonly used to *suppress* the NOE while still providing the spectral simplification afforded by decoupling. Rather than leaving the decoupling r.f. source engaged continuously, it is switched (gated) to operate for the minimum time necessary to achieve decoupling, i.e. *only* during the actual acquisition of the spectrum and *not* during any delays between successive acquisitions.

4.2 Molecular geometry using NOE

The NOE is a powerful method for establishing the three-dimensional proximity of nuclei (or groups of nuclei) in space and is used extensively in mapping and establishing the conformations, configurations and stereochemistry of molecules in solution. Since the nuclear Overhauser effect is a result of dipole–dipole interactions between nuclei, the magnitude of the NOE depends on the distance between the interacting nuclei. The NOE diminishes dramatically (as the sixth power) as the distance between saturated and observed nuclei (r) increases:

$$\text{NOE enhancement factor} \propto \frac{\gamma_{sat}\gamma_{obs}}{r^6} \qquad (9)$$

5   THE NMR EXPERIMENT AND INSTRUMENTATION

An important consideration for any analytical measurement is the ability to identify and quantify reliably some desired signal in the presence of background noise arising primarily from the electronics of the detector. The sensitivity of an NMR spectrometer is conventionally quantified in terms of the signal-to-noise ratio (usually abbreviated to S/N), defined as the ratio of the peak height (H) of a signal divided by the maximum peak-to-peak amplitude of the noise (N) divided by 2.5:

$$\text{S/N} = \frac{H}{N/2.5} \qquad (10)$$

For testing and comparing the sensitivity of spectrometers, standard solutions of known composition are usually employed. ^1H sensitivity is routinely assessed using a single scan of the CH_2 resonance in a sample of ethylbenzene in $CDCl_3$ solution (0.1 % v/v) in a 5 mm o.d. NMR tube. ^{13}C sensitivity is usually assessed using the tallest aromatic resonance (single scan, broadband ^1H decoupled) in a sample of ethylbenzene in $CDCl_3$ (10% v/v) or alternatively the aromatic carbon resonance (single scan, no ^1H decoupling) in a solution of deuterio-benzene–dioxane (60:40 v/v).

The S/N is a function of many variables, some of which are instrumental and some related to the nature of the sample. Among the main determinants of S/N are: (i) the strength of the magnetic field, S/N $\propto (\mathbf{B_0})^{3/2}$; the more intense the magnetic field the better will be the S/N achievable by the spectrometer; (ii) the design of the r.f. probe, in particular the length and geometry of the receiver coil; and (iii) the tuning of the resonant frequency and matching the impedence of the probe to that transmitter and receiver; samples with different dielectric constants 'detune' the NMR probe cavity to different extents so the adjustment of the impedence and frequency of the probe should be performed for each sample to achieve the optimum S/N.

There are two distinct types of commercially available NMR spectrometers: continuous wave (CW) and pulsed or Fourier Transform (FT) spectrometers. CW instruments were historically the first to be developed commercially, but most modern spectrometers are of the FT type. Both types of spectrometer require a strong, homogeneous magnet, an r.f. source and a detector. The instruments perform essentially the same experiment but the operating hardware and mode of performing the measurements are entirely different.

5.1 CW NMR spectrometers

The CW NMR spectrometer detects the NMR frequencies of nuclei in a sample placed in a magnetic field by sweeping the frequency of r.f. radiation through a given range and directly recording the intensity of absorption as a function of frequency. In practice it takes several minutes (typically 1–10 min) to scan through the desired range of frequencies for a particular type of nucleus and then record the spectrum. The spectrum is usually recorded and plotted simultaneously with a recorder synchronized to the frequency of the r.f. source. Although it is possible to digitize and store successive scans through a given frequency region (and hence accumulate signal intensity), this is extremely time consuming in practice, and has been done only rarely.

The primary advantage of the CW spectrometer is that the initial cost is relatively low. The CW spectrometer has little high-technology hardware (hence maintenance is minimal). CW instruments are easy to operate and most small (low-field) laboratory spectrometers for routine applications are still of the CW type.

5.2 Pulsed FT NMR spectrometers

The pulsed or Fourier transform NMR spectrometer employs a single high-power, short-duration pulse of r.f. radiation to excite simultaneously all nuclei whose resonance frequency is near the frequency of the pulse. Immediately following the r.f. pulse, the sample radiates a signal, termed a free induction decay (FID), which is modulated by the frequencies of all nuclei excited by the pulse. The signal detected as the nuclei return to equilibrium (intensity as a function of time) is recorded, digitized and stored as an array of numbers in a computer. Fourier transformation of the data affords a conventional (intensity as a function of frequency) representation of the spectrum (Figure 4).

Typically, an FID may range from a few microseconds to tens of seconds in duration. The duration of the FID depends on the T_1 relaxation times of the nuclei in the sample and on factors, including the uniformity of the magnetic field across the sample volume (homogeneity). Since the frequencies of all nuclei are detected simultaneously in a single FID, the time for data acquisition in an FT experiment is much shorter than in the corresponding CW experiment. The speed with which an experiment can be repeated is determined by the relaxation

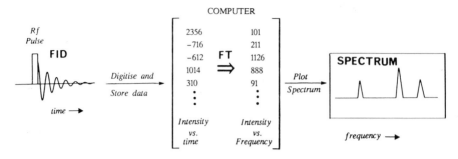

Fig. 4. Sequence of steps involved in obtaining a pulsed NMR spectrum

times of nuclei in the sample and, because the data are captured in digital form, the coaddition of FIDs is not a difficult process. Many FIDs can be acquired in the time taken to accumulate a single CW scan. The advent of FT spectrometers permitted the routine investigation of important nuclei with inherently weak NMR signals (e.g. ^{13}C) which are essentially undetectable by CW methods.

The FT NMR spectrometer is a more advanced and sophisticated instrument than the CW type. Instruments for pulsed FT NMR spectroscopy must be equipped with a computer with the capability to control the r.f. pulse precisely, digitize and capture the FID rapidly, and perform the mathematics of the Fourier transformation. The necessity for a computer and the associated interfacing and data-storage requirements makes the FT spectrometer signifi- cantly more expensive than its CW counterpart. The complexity of the instru- ment and the multi-stage nature of the experiment dictate that the instrument operator needs to be more highly trained than for CW instrumentation. FT NMR spectrometers not only permit the same experiments as CW instruments (but more efficiently), they also enable an ever-expanding range of pulsed experiments which are not accessible by the CW approach to be performed.

5.2.1   The pulse angle

As described above (Section 3), when placed in a magnetic field a sample is magnetized (slightly) in the direction of the field. The absorption of r.f. radiation by the nuclei induces magnetization (termed *transverse magnetization*) in the plane perpendicular to B_0 and it is this magnetism which is detected by the receiver coil of an NMR spectrometer.

The power of an r.f. pulse is usually expressed in units of degrees or radians. This unit arises because the intensity of the signal detected in pulsed NMR experiment is a sinusoidal function of the power of the r.f. pulse employed for excitation (Figure 5). A 90° pulse (alternatively termed a $\pi/2$ pulse) is the power required to obtain a maximum response from the sample. A 90° pulse is sufficient power exactly to equalize the populations of the upper and lower

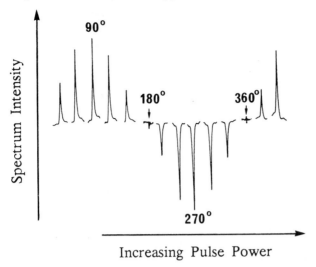

Fig. 5. Variation in the intensity of the ^1H spectrum of $CHCl_3$ as a function of the power of the r.f. pulse

energy states of a spin system (see Fig. 2). Doubling the power of a 90° pulse (i.e. a 180° or π pulse) produces no signal from the sample; a 180° pulse inverts the populations of the upper and lower energy states such that the α state is more highly populated than the β state. A 450° ($5\pi/2$) pulse will give an identical response to the 90° pulse and a 45° ($\pi/4$) pulse will give a signal of 0.707 (i.e. sin 45) the intensity of the response following a 90° pulse.

For a given sample, the S/N is improved by adding together spectra or FIDs. Noise is phase incoherent (i.e. its phase varies randomly from one FID to another) whereas as 'real' NMR signals are coherent. Real signals always add constructively whereas noise does not and, in quantitative terms, the S/N increases as the square root of the number of FIDs or spectra added.

5.2.2 Free induction decay (FID)

The FID which is obtained from a sample following an r.f. pulse is the superposition of signals arising from all nuclei excited by the pulse. The rate of exponential decay of signals in the FID is related to the T_1 of the nuclei in the sample and in non-viscous liquids, the decay rate of the FID is essentially equal to T_1. However, other factors including the homogeneity and stability of the magnetic field, the homogeneity of the sample (i.e. the presence of particulate material in a solution), chemical exchange of nuclei and the presence of paramagnetic species in the sample also influence the length of time for which transverse magnetization persists in a sample and hence the time for which a

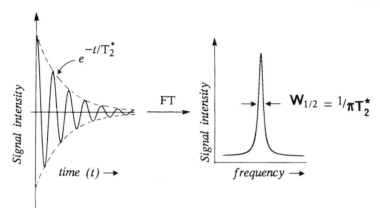

Fig. 6. Fourier transformation of an FID which decays exponentially with a time constant of T_2^* s gives rise to a Lorentzian lineshape whose width at half-height is $1/\pi T_2^*$ Hz

signal can be detected. The decay rate of the FID can be significantly more rapid than that dictated by the T_1 relaxation time of nuclei in the sample. The experimentally observed rate of decay of the FID is termed T_2^* and it is always less than (or equal to) T_1, i.e. even though the FID has decayed essentially to zero, the populations of nuclei between the various available states may take significantly longer to return to equilibrium. Particularly for solid samples, T_2^* can be very short whereas T_1 may be very much longer.

The Fourier transform of an exponentially decaying signal results in a Lorentzian lineshape. If the time constant for the exponential decay is T_2^* then the Lorentzian line will have a width at half-height ($W_{1/2}$) in hertz given by $W_{1/2} = 1/\pi T_2^*$ (Figure 6).

It can be seen that a rapid decay of signals which make up the FID results in broad lines in the NMR spectrum. In any NMR spectrum, the nuclei which contribute signals to the FID have different relaxation times. Nuclei which have short relaxation times (and hence give rise to broad lines in the NMR spectrum) will contribute most significantly to the beginning of the FID. On the other hand, nuclei with long relaxation times contribute to the entire FID and hence give rise to narrow lines in the NMR spectrum.

The acquisition time.

To resolve two resonances separated by Δv Hz, the FID must be acquired for at least $1/\Delta v$ s. Conversely, if the FID decays such that it cannot be measured after t s, one cannot resolve spectral features separated by $1/t$ Hz. Although an exponentially decaying signal persists for infinite time, in practice the background noise level of the NMR receiver swamps the small amount of signal in the FID after the intensity has decayed significantly. The noise is, of course, present throughout the entire period of detection of the FID, but at the beginning of the FID the signal from the sample is usually large compared with the noise, whereas at the end the noise outweighs the signal. Once the signal has decayed to a level well below the instrumental noise there is no reason to continue to acquire the FID since there is no real contribution to the required NMR spectrum beyond this point. In practice, the acquisition time should be 3–4 times T_2^*.

The rate of data sampling.

Following an r.f. pulse, the FID must be digitized and stored to permit Fourier transformation. The intensity of the NMR signal is sampled at equal intervals, converted to numerical form (with an analogue-to-digital converter) and stored as an array of integers. The interval (Δt) between points at which the FID is sampled establishes the highest frequency which can be measured in the FID and this defines the detected spectral width (SW) of the FT NMR experiment:

$$SW = \frac{1}{2\Delta t} \tag{11}$$

For the observation of large spectral widths (as is often necessary with the spectra of solids), the rate of digitization must be rapid.

The repetition rate (recycle time).

In order to achieve the best S/N in a given period of time, there are obviously a number of factors which need to be balanced. Although the maximum signal is obtained from a sample following a $\pi/2$ pulse, one must wait for several T_1 periods between successive pulses to allow recovery of the population. If the recycle time is too short and population recovery is not complete, the signal available for each successive acquisition is diminished. If the pulse angle is less than $\pi/2$, the observed signal in each acquisition is diminished but the nuclei in the sample takes less time to recover between successive pulses.

The optimum value of the pulse angle (termed the Ernst angle, θ_E) for a given experiment repetition rate (t_r) is given by

$$\cos \theta_E = \exp(-t_r/T_1) \tag{12}$$

5.2.3   Data massage and weighting

Once the FID has been converted into digital form (i.e. as an array of numbers) and stored in a computer, there is an opportunity to take the data and massage or mathematically modify it to either improve the S/N of the spectrum, improve the resolution of the spectrum, select out specific information or filter out 'unwanted' information. For many experiments, data massage can increase the capacity of the spectroscopist to obtain useful data from the spectrum. It must be emphasized that any form of data massage means that the spectrum is no longer a faithful representation of the composition of the sample. Data which have been weighted are biased in some fashion and must be analysed with a complete knowledge of the consequences of the 'massage'.

Two of the most useful and widely used massaging processes are those involved in either improving the S/N of a spectrum or enhancing the signal resolution.

Enhancement of the S/N can be achieved easily by weighting the FID in favour of the data at the beginning. Typically this is achieved by multiplying the data in the FID by a function which decreases exponentially with time. This has the effect of amplifying data at the beginning of the FID and reducing the data at the end. Since noise contributes most significantly in the later parts of the FID, the noise is reduced in the spectrum (see Figure 7b). However, this type of weighting also suppresses the signals from narrow lines (i.e. those with long T_1 values) so the FID (and the spectrum) is biased against signals with narrow lines and consequently the spectrum 'resolution' is degraded.

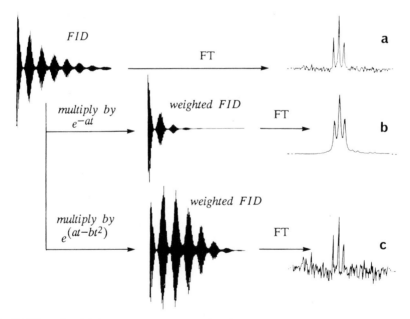

Fig. 7. Effect of weighting an FID. (a) No weighting of the FID; (b) multiplication by an exponentially decreasing function to improve the S/N; (c) multiplication by a Gaussian function to enhance the resolution

Resolution enhancement can be achieved by multiplying the data in the FID by a function which accentuates data at the end of the FID. Multiplication by Gaussian or exponentially increasing functions is commonly employed to amplify data in the later parts of the FID while decreasing the relative contribution of the data at the beginning (see Figure 7c). Broad lines (whose signals decay rapidly) are suppressed while the contribution from nuclei with long relaxation times (i.e. narrow lines) is enhanced. In addition to improving the overall resolution of the signals in the spectrum, this method of weighting also enhances the noise.

6 ADVANCED CONCEPTS

6.1 Pulse sequences

In a pulsed NMR experiment, the FID acquired is the signal from the sample after the nuclei have been excited by an r.f. pulse. The pulsed NMR experiment can be divided into three parts, usually termed the relaxation, preparation and acquisition periods (Figure 8). In most applications, the entire process is repeated many times and the data are summed to improve the S/N. A single r.f. pulse is the simplest means of preparing or exciting the nuclei, but combinations of pulses also generate a detectable signal from the sample.

A programmed sequence of r.f. pulses, separated by defined intervals, can be designed such that particular properties of the sample or its spin system can be examined. Using specific pulse sequences, the FID (and hence the spectrum obtained after Fourier transformation may be enhanced, modulated or encoded with some desired property of the sample. In most commercially available NMR spectrometers, it is possible to control the frequency, phase and power of the r.f.

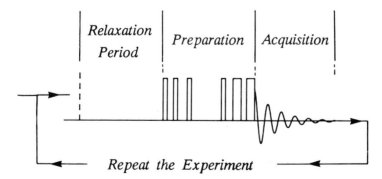

Fig. 8. The pulse NMR experiment. The preparation period can be a single pulse or a sequence of pulses

pulses in a programmed sequence in addition to the separation between pulses. There are an unlimited number of pulse sequences possible, and specific sequences have been designed, for example: to differentiate or measure the relaxation rates of the various nuclei in the sample; to detect the presence of spin–spin coupling between pairs of nuclei; to suppress contributions from dipolar coupling in the spectra of solids; to suppress large signals (e.g. from solvents) which otherwise make regions of the spectrum unusable; to assess spatial proximity between pairs of nuclei; to detect and quantify exchange process in the sample, e.g. groups or atoms which move from one environment to another; and to suppress and correct artefacts which arise from imperfections in the NMR spectrometer hardware.

6.1.1   Inversion–recovery measurement of T_1

The T_1 values for nuclei in a sample are most commonly measured using a simple two-pulse sequence given in Figure 9. (For a discussion of experimental methods for measuring T_1, see Ref. 1b, Ch. 7, pp. 244–290, and references cited therein.) A single π pulse is used to disturb (in this case invert) the equilibrium populations of the energy levels of nuclei in the sample. Following the pulse, the populations begin to recover towards equilibrium during the delay τ. A $\pi/2$ pulse is used to convert the population difference between levels to a detectable

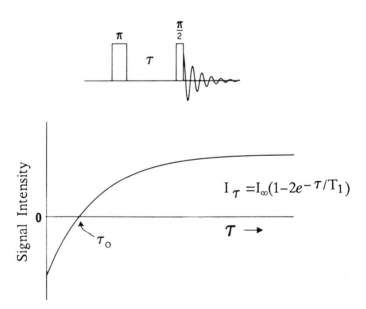

Fig. 9. The pulse sequence employed in inversion–recovery measurement of T_1. The experiment is repeated a number of times and the intensity of the observed spectrum is an exponential function of τ

NMR signal. When τ is zero (or very small) the sample effectively receives a $3\pi/2$ pulse and the observed signal is negative. When τ is long, the sample relaxes fully between the π and $\pi/2$ pulses and maximum (positive) signal (I_∞) is observed. Between these extremes, the observed signal (I_τ) increases exponentially with τ (Figure 9). T_1 is most conveniently obtained from the slope of a graph of $\ln(I_\infty - I_\tau)$ against τ.

When $\tau = 0.693 T_1$, no NMR signal is observable and the value of τ where the NMR signal is nulled (τ_0 in Figure 9) can be used to estimate T_1. Alternatively, by setting τ to the value where the observable signal is exactly zero, an unwanted signal (for example a solvent resonance[10]) can be effectively removed from a spectrum and signals from nuclei with different T_1 values can be observed.

6.1.2 INEPT and DEPT

INEPT (*I*nsensitive *N*uclei *E*nhanced by *P*olarization *T*ransfer)[11] and DEPT (*D*istortionless *E*nhancement by *P*olarization *T*ransfer)[12] are heteronuclear pulse sequences which involve the simultaneous application of a programmed sequence of pulses to the proton spectrum and to the spectrum of a hetero-nucleus X (e.g ^{13}C) (Figure 10). Both INEPT and DEPT provide an enhancement of the intensity of the signals from protonated atoms in the heteronuclear spectrum by a factor of γ_H/γ_X (i.e. a factor of 3.98 for ^{13}C).

By a judicious choice of the delay τ_2 in INEPT or the pulse angle θ in the DEPT pulse sequence, the experiments can be employed to change selectively the amplitude and/or phase of the signals in the X spectrum belonging to XH groups, XH_2 groups or XH_3 groups[13]. For example, with τ_2 set to $0.75/{}^1J_{XH}$, the INEPT experiment gives CH_3 and CH resonances upward and CH_2 resonances inverted in the ^{13}C NMR spectrum.

Heteronuclear spectra can be 'edited' to generate sub-spectra containing, for example, only signals from XH groups or only signals from XH_2 groups, etc., by taking the sums and differences of spectra which have XH, XH_2 and XH_3 groups phased differently. Most modern FT spectrometers have programs to execute automatically the necessary spectral acquisitions and to perform the appropriate additions and subtractions to provide edited sub-spectra routinely.

6.1.3 Hartmann–Hahn cross-polarization

Cross-polarization[14] is a technique commonly employed in obtaining the ^{13}C spectra from solid samples (Figure 11). The T_1 relaxation times of ^{13}C nuclei in the solid state can be prohibitively long whereas the T_1 values for ^{1}H are typically much smaller. Cross-polarization is a method by which the ^{13}C population difference is established by transfer from protons. The rate of ^{1}H relaxation (and not the rate of ^{13}C relaxation) thus dictates the rate at which ^{13}C acquisition can be repeated. Since the population difference between levels

INEPT

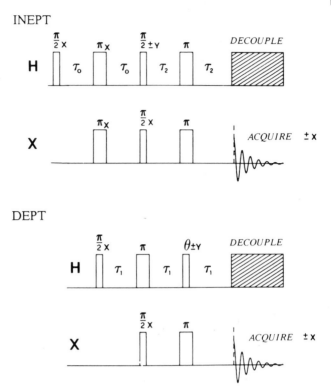

Fig. 10. Pulse sequences employed for the INEPT and DEPT experiments (pulse phases not indicated). The delays τ_0 and τ_1 are set $\tau_0 = 0.25/^1J_{XH}$ and $\tau_2 = 0.5/^1J_{XH}$. τ_2 and θ are explained in the text

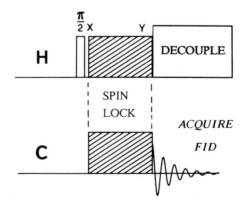

Fig. 11. Basic pulse sequence employed for Hartmann–Hahn cross-polarization in solid-state NMR spectroscopy

in the proton spectrum is larger than for the ^{13}C spectrum, an enhancement of the ^{13}C spectrum is obtained by population transfer from ^{1}H (by a factor of ca 4).

The period labelled 'spin lock' in the pulse sequence is typically several milliseconds in duration and in this period both ^{13}C and the ^{1}H spectra are irradiated. Interaction of the protons with the r.f. field causes the protons to align and precess about the rotating field of the r.f. radiation; similarly, interaction of ^{13}C nuclei with irradiation at the ^{13}C frequency aligns them with the rotating field of the r.f. radiation. The strength of the irradiating ^{13}C r.f. field (H_c) and the irradiating ^{1}H r.f. field (H_H) must be adjusted such that $\gamma_H H_H = \gamma_C H_C$ (termed the Hartmann–Hahn condition). When this condition is met, the ^{1}H and ^{13}C nuclei precess with the same frequency with respect to the spin-locking field and during the spin-locked period (the 'contact time') equilibration of the proton and ^{13}C populations proceeds by population transfer from the ^{1}H spectrum to the ^{13}C spectrum. The quantity $T_{1\rho}$ is defined as the relaxation time in the rotating frame (along the direction of the spin-locking field) and is an important characteristic property which is sensitive to molecular motion (see Chapter 5, Section 2.3).

6.1.4 WAHUHA

The WAHUHA[15] (*Waugh–Huber–Haeberlen*) pulse sequence and its variations[16] are line-narrowing sequences used to reduce the observed width of resonances in the NMR spectra of solids. The sequence (Figure 12) averages dipolar couplings to zero and reduces chemical shifts. For the ^{1}H spectroscopy of solids, linewidths can be reduced from several kilohertz to 100–200 Hz by application of the WAHUHA sequence.

In the WAHUHA sequence, the cycle of four pulses is repeated throughout the acquisition of the FID with one data point of the FID sampled during each of the 2τ delays. The interval between points in the FID (and hence the spectral width, see Section 5.2.2) is governed by the selection of τ.

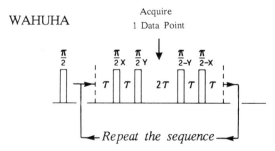

Fig. 12. The basic WAHUHA pulse sequence

6.1.5   WALTZ

WALTZ[17] is a pulse sequence designed for broadband heteronuclear decoupling. Decoupling is achieved not by the application of continuous r.f. irradiation but by application of a train of phase-cycled pulses to the nuclei being decoupled.

The WALTZ pulse sequence or its variations[18] decouple over a much larger bandwidth than conventional broadband decoupling methods and even in high-frequency spectrometers (where the demands on bandwidth are greatest) efficient decoupling can be achieved with a decoupler power which does not cause significant sample heating.

6.2   Two-dimensional NMR spectroscopy

In spectra which are inherently crowded, e.g. those from complex mixtures of different compounds or the spectra of polymers, overlapping resonances can often be separated or dispersed by introducing an additional dimension to the normal intensity vs frequency representation of an NMR spectrum.

A two-dimensional NMR spectrum[19] is acquired using a pulse sequence which contains some delay period t_1 which can be varied systematically as the experiment is repeated. The experiment is repeated many times (typically 512 or 1024), with a different time t_1 in the pulse sequence for each experiment. One FID is acquired for each experiment, giving an array of N individual FIDs, each of which has been acquired with a slightly different pulse sequence. Each FID represents the variation of detected signal as a function of time (t_2 in Figure 13) and successive FIDs in the array differ as a function of the time variable t_1 within the preparation period of the pulse sequence.

Fourier transformation of the two-dimensional array of data with respect to t_2 affords a series of spectra which might vary in intensity and phase as a function of t_1. A second Fourier transformation, this time with respect to t_1, gives a two-dimensional spectral array (which is function of frequency domains f_1 and f_2). Two-dimensional spectra are usually represented in terms of a stacked plot or contour plot. The contour plot (see, for example, Figure 14) is a more convenient representation for making measurements or peak assignment.

There are an unlimited number of two-dimensional experiments which differ simply in the specific nature of the pulse sequence employed during the preparation period. There are several commonly used two-dimensional experiments which deserve some description.

6.2.1   COSY

The COSY[20] (Shift Correlated Spectroscopy) experiment uses a pulse sequence which modulates the resonance of each nucleus in an NMR spectrum with the

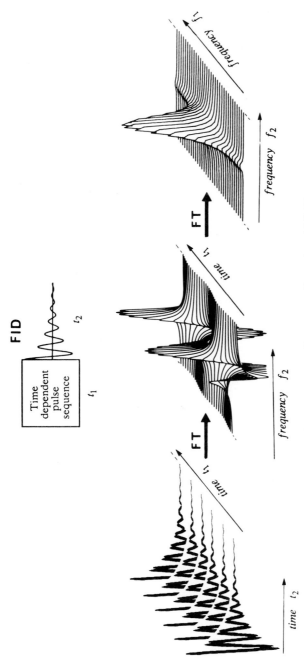

Fig. 13. Stages involved in recording a two-dimensional NMR spectrum

chemical shift of nuclei which are coupled to it. The basic pulse sequence (Figure 14) consists of two pulses separated by a variable delay. The two-dimensional spectrum obtained after FT contains signals along one diagonal corresponding to the frequencies of nuclei in the sample and off-diagonal peaks (cross-peaks) which indicate the presence of spin–spin coupling between pairs of nuclei. In *one* experiment, the presence or absence of spin–spin coupling between *all* pairs of nuclei in the sample can be established. This technique is particularly important for examining large biological molecules or polymers where both the extremely large number of resonances and the severe overlap of signals makes stepwise (one-dimensional) selective decoupling experiments impossible.

The heteronuclear analogue of the COSY pulse sequence termed heteronuclear shift correlation (HSC) provides a correlation between the chemical shifts of the protons and the chemical shifts of heteronuclei which are coupled to

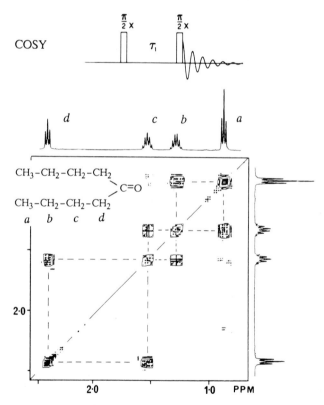

Fig. 14. Contour plot of a ^{1}H COSY spectrum of dibutyl ketone. The presence of cross-peaks indicates the sequential coupling path from H_a to H_b, H_b to H_c and from H_c to H_d. Normal one-dimensional spectra are included on the vertical and horizontal axes to indicate the resonance positions

them. This technique is useful in identifying and assigning resonances because the known peak assignments of the proton spectrum can be used to assign directly resonances of carbon atoms (or other heteroatoms) which show spin–spin coupling to protons or *vice versa*.

6.2.2   *J*-resolved spectroscopy[21]

This technique separates chemical shift and coupling information so that resonances are split only by spin–spin coupling in one dimension and only by chemical shift in the other. This is a useful method for unravelling crowded spectra with overlapping multiplets.

6.2.3   NOESY[22]

This is a two-dimensional NOE experiment. The basic pulse sequence (Figure 15) consists of three pulses and two delays, one of which is variable (t_1) and the other constant (τ_m). The spectrum obtained is superficially similar to that of a COSY in appearance, but the pulse sequence employed gives rise to cross-peaks only when two nuclei are close together in space.

NOESY

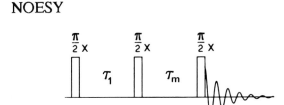

Fig. 15. The pulse sequence employed for the two-dimensional NOESY experiment

7   CONCLUSION

The individual chapters which follow address specialist areas of the analytical application of NMR spectroscopy. Each chapter deals in depth with a single aspect of NMR spectroscopy, but each assumes a level of fundamental understanding of the subject. Certain fundamental topics and experiments have necessarily been dealt with in a superficial manner and the reader should consult the books and/or articles referenced, particularly in Ref. 1 here, in instances where a more detailed description is required.

REFERENCES

1. For general references on NMR spectroscopy, see for example (a) M. L. Martin, J.-J. Delpuech and G. J. Martin, *Practical NMR Spectroscopy*, Heyden, London, 1980; (b) A. E. Derome, *Modern NMR Techniques for Chemistry Research*, Pergamon Press, Oxford, 1987; (c) D. Shaw, *Fourier Transform NMR Spectroscopy*, Elsevier, Amsterdam, 1984; (d) J. K. M. Saunders and B. K. Hunter, *Modern NMR Spectroscopy*, Oxford University Press, Oxford, 1987; (e) H. Günther, *NMR Spectroscopy*, Wiley, New York, 1980.
2. See for example (a) R. K. Harris and B. E. Mann (eds), *NMR and the Periodic Table*, Academic Press, London, 1978; (b) C. Brevard and P. Granger, *Handbook of High Resolution Multinuclear NMR*, Wiley, New York, 1981.
3. For a detailed description of the fundamental NMR parameters, see for example (a) C. P. Slichter, *Principles of Magnetic Resonance*, Springer, Berlin, 1980; (b) A. Abragam, *The Principles of Nuclear Magnetism*, Oxford University Press, London, 1961; (c) R. R. Ernst, G. Bodenhausen and A. Wokaun, *Principles of Nuclear Magnetic Resonance in One and Two Dimensions*, Oxford University Press, Oxford, 1987.
4. See for example (a) J. W. Emsley (ed.), *Nuclear Magnetic Resonance of Liquid Crystals*, Reidel, Dordrecht, 1985; (b) J. W. Emsley and J. C. Lindon, *NMR Spectroscopy Using Liquid Crystal Solvents*, Pergamon Press, Oxford 1975; (c) P. Diehl and C. L. Khetrapal. In P. Diehl, E. Fluck and R. Kosfeld (eds), *NMR—Basic Principles and Progress*, Vol 1, Springer, Berlin, 1969, pp. 3–93; (d) C. L. Khetrapal, A. C. Kunwar, A. S. Tracey and P. Diehl. In P. Diehl, E. Fluck and R. Kosfeld (eds), *NMR—Basic Principles and Progress*, Vol. 9, Springer, Berlin, 1975, pp. 3–74.
5. A. Pines, M. G. Gibby and J. S. Waugh, *J. Chem. Phys.*, **54**, 569 (1973).
6. E. R. Andrew, *Philos, Trans. Roy. Soc. London, Ser. A*, **299**, 505 (1981).
7. J. H. Noggle and R. E. Schirmer, *The Nuclear Overhauser Effect—Chemical Applications*, Academic Press, New York, 1972.
8. Reference 1b, Ch. 5, pp. 97–128.
9. Reference 1c, Ch. 6, pp. 163–205.
10. P. J. Hore, *J. Magn. Reson.*, **55**, 283, (1983), and references cited therein.
11. (a) G. A. Morris and R. Freeman, *J. Am. Chem. Soc.*, **101**, 760 (1979); (b) D. P. Burum and R. R. Ernst, *J. Magn. Reson.*, **39**, 163 (1980); (c) A. J. Shaka and R. Freeman, *J. Magn. Reson.*, **50**, 502 (1982).
12. (a) D. M. Doddrell, D. T. Pegg and M. R. Bendall, *J. Magn. Reson.*, **48**, 323 (1982); (b) D. T. Pegg, D. M. Doddrell and M. R. Bendall, *J. Chem. Phys.*, **77**, 2745 (1982); (c) O. W. Sorenson and R. R. Ernst, *J. Magn. Reson.*, **51**, 477 (1983); (d) G. A. Morris, *Top. ^{13}C NMR Spectrosc.*, **4**, 179 (1984), and references cited therein.
13. (a) D. T. Pegg, D. M. Doddrell, W. M. Brooks and M. R. Bendall, *J. Magn. Reson.*, **44**, 32 (1981); (b) J. M. Bulsing, W. M. Brooks, J. Field and D. M. Doddrell, *J. Magn. Reson.*, **56**, 167 (1984); (c) M. R. Bendall and D. T. Pegg, *J. Magn. Reson.*, **53**, 272 (1983); (d) O. W. Sorensen, S. Bildsoe, H. Bildsoe and H. J. Jacobsen, *J. Magn. Reson.*, **55**, 347 (1983); (f) D. T. Pegg and M. R. Bendall, *J. Magn. Reson.*, **60**, 347 (1984).
14. A. Pines, M. G. Gibby and J. S. Waugh, *J. Chem. Phys.*, **59**, 569 (1973).
15. (a) J. S. Waugh, L. M. Huber and U. Haeberlen, *Phys. Rev. Lett.*, **20**, 180 (1968); (b) U. Haeberlen and J. S. Waugh, *Phys. Rev.*, **175**, 453 (1968).
16. See for example (a) B. C. Gerstein, R. G. Pembleton, R. C. Wilson and C. M. Ryan, *J. Chem. Phys.*, **66**, 361 (1977); (b) L. M. Ryan, R. E. Taylor, A. J. Paff and B. C. Gerstein, *J. Chem. Phys.*, **72**, 508 (1980); (c) W. K. Rhim and D. P. Burum, *J. Chem.*

Phys., **71**, 944 (1978); (d) B. C. Gerstein, *Philos. Trans. Roy. Soc. London, Ser. A*, **299**, 521 (1981)

17. (a) A. J. Shaka, J. Keeler and R. Freeman, *J. Magn. Reson.*, **53**, 313 (1983); (b) M. H. Levitt, R. Freeman and T. Frankiel, *J. Magn. Reson.*, **47**, 328. (1982).
18. See for example A. J. Shaka, P. B. Barker and R. Freeman, *J. Magn. Reson.*, **64**, 547 (1985).
19. For general descriptions of 2D NMR and common 2D experiments, see for example (a) Reference 1d Ch. 4–8; (b) A. Bax, *Two-Dimensional Nuclear Magnetic Resonance in Liquids*, Delft University Press, Delft, 1982; (c) Reference 1b, Ch. 8–10; (d) R. Benn and H. Günther, *Angew. Chem., Int. Ed. Engl.*, **22**, 350 (1983); (e) H. Kessler, M. Gehrke and C. Griesinger, *Angew. Chem., Int. Ed. Engl.*, **27**, 490 (1988).
20. (a) W. P. Aue, E. Bartholdi and R. R. Ernst, *J. Chem. Phys.*, **64**, 2229 (1976); (b) G. Bodenhausen, H. Kogler and R. R. Ernst, *J. Magn. Reson.*, **58**, 370 (1984).
21. (a) W. P. Aue, J. Karhan and R. R. Ernst, *J. Chem. Phys.*, **64**, 4226 (1976); (b) L. Müller, A. Kumar and R. R. Ernst, *J. Chem. Phys.*, **63**, 5490 (1975).
22. (a) J. Jeener, B. H. Meier, P. Bachmann and R. R. Ernst, *J. Chem. Phys.*, **71**, 4546 (1979); (b) G. Bodenhausen H. Kogler and R. R. Ernst, *J. Magn. Reson.*, **58**, 370 (1984); (c) K. Wüthrich, *NMR of Proteins and Nucleic Acids*, Wiley, New York, 1986.

Analytical NMR
Edited by L. D. Field and S. Sternhell
© 1989 John Wiley & Sons Ltd

Chapter 3

Quantitative Applications of ^{13}C NMR

J. R. Mooney

Standard Oil Co., Warrenville Center, Cleveland, OH 44128, USA

1 BACKGROUND

Although several excellent review articles have addressed different aspects of quantitative NMR[1,2], previous textbooks have not dealt thoroughly with the issue. Perhaps the major reason is the fact that a certain belief has built up that pulsed FT NMR spectrometers are not quantitative tools. This is far from the true situation—it is possible not only to do adequate quantitative spectroscopy at the percentage level but also to do quantitation at the parts per thousand level with acceptable precision and accuracy.

In this chapter, the examples used and the techniques proposed are primarily based on ^{13}C NMR studies. The techniques are directly applicable to any other nucleus where relaxation problems and nuclear Overhauser enhancement (NOE) problems exist. In quantitative experiments where only one type of nucleus need be considered, e.g. ^{1}H or ^{19}F, the techniques for NOE suppression can be ignored. The chapter is divided into three sections for convenience: (i) NMR in solution, (ii) NMR in solids and (iii) practical considerations.

The primary problem that has limited the development of good quantitative ^{13}C NMR practices is the assumed lack of sensitivity of the ^{13}C nucleus. The limited sensitivity established the practice of taking spectra under maximum sensitivity conditions, i.e. acquisition using the Ernst angle (see Chapter 2, Section 5.2.2). With today's modern high-field, high-signal-to-noise spectrometers, it is no longer necessary to consider only qualitative spectra. All current spectrometers give more than adequate sensitivity using a single pulse to measure accurately parts per hundred level quantities on the ASTM signal-to-noise reference sample (the ASTM reference sample is 60:40 benzene-d_6-p-dioxane; typical one-pulse spectra measure in excess of 500:1 S/N). Therefore, if it is possible to prepare a similar sample and run it under similar conditions, it is possible to do quantitative work routinely.

Nevertheless, substantial difficulties remain in selecting the correct experimental conditions to perform highly precise quantitative analysis of ^{13}C and other nuclei. The two phenomena that determine how the experiment can be performed are the T_1 relaxation time and the NOE factor. The effect of T_1 dominates all quantitative measurements and, for most carbon species in solution, dipole–dipole relaxation is the dominant contributor to T_1. For small molecules and end-groups in larger molecules, spin–rotation relaxation can also be significant and, for those molecules containing anisotropic groups, chemical shift anisotropy can be particularly important. In solids and in certain solutions, paramagnetic (electron–dipole) relaxation can dominate.

Because it is desirable to decouple ^{1}H while acquiring ^{13}C spectra to allow simplicity of interpretation, the nuclear Overhauser effect must be taken into account. As described in Chapter 2, normal proton decoupling provides an enhancement of the ^{13}C spectrum. The size and extent of the NOE are dependent on the dipole–dipole interaction between ^{1}H and ^{13}C nuclei. Therefore, in cases where dipole–dipole relaxation is important, we expect a maximum NOE factor, η of 1.988. As spin–rotation or paramagnetism contributes more to relaxation, η will be decreased.

$$\eta = \frac{1.988 \, T_1^{\text{other}}}{T_1^{\text{DD}} + T_1^{\text{other}}}$$

Within any given sample, the large variations in T_1 and NOE require that we find some way of equalizing or eliminating them. One method of dealing with NOE is simply to acquire spectra without proton decoupling. Early workers successfully used this method for determining the aromaticity of petroleum hydrocarbons. However, this is not a generally useful technique as we lose the high level of specificity of the ^{13}C method. The more commonly used practice is to turn off the decoupler during the time required for the nuclei to relax. This method, normally referred to as 'inverse gated decoupling', has been widely used. By calibrating against an internal standard, one can correct for NOE differences. In some cases, such as in the analsyis of mixtures of stereoismers, one can simply ignore NOE differences.

Many techniques for dealing with the difficulties encountered in doing reliable quantitation have been developed. Procedures differ for small molecules, polymers and solids. All have the common feature of establishing the spin equilibrium of the rate-determining ^{13}C relaxation time. Normally, this will require running spectra such that the recycle time between acquisitions is 5–7 times the T_1 of interest.

2 ^{13}C NMR IN SOLUTION

2.1   Optimizing recycle time for acquisition

For small molecules in solution, where it is desired to obtain correct relative intensities of different carbons in a molecule (i.e. organic structure proof), it is necessary to suppress the NOE. For this type of work, inverse gated decoupling (decouple only during acquistion) is the common practice.

The longest relaxing ^{13}C nucleus, invariably a quaternary carbon, becomes the carbon that establishes the rate of the acquisition. One of the simplest examples that we can consider is how to acquire a fully equilibrated spectrum of a simple molecule. Gillet and Delpuech[3] examined benzene, both experimentally and theoretically. Benzene has $T_{1C} = 29.5$ s, $T_{1H} = 17.5$ s and NOE $= 1.60$[3]. The results of applying different pulse–delay conditions are shown in Figure 1 and these results can be understood in terms of the inverse gated decoupled experiment. During acquisition (set at 1.2 s) with proton decoupling, NOE builds as a function of T_1. NOE decays along with spin polarization during the delay (t_3). Using 30° and 60° pulse excitation, it is clear that there are values of t_3 where NOE buildup compensates for the effect of incomplete relaxation between acquisitions. Hence there is a case for this simple molecule (and every T_{1C}, T_{1H} combination) where maximum magnetization can be achieved under

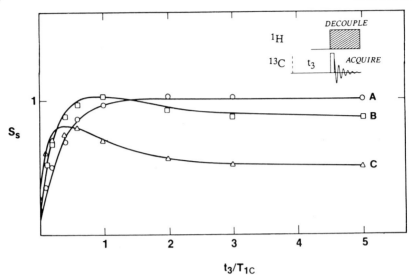

Fig. 1. Plot of the steady-state signal intensity for benezene as a function of the ratio t_3/T_{1C}. Theoretical curves, A, B, C are represented by solid lines for pulse widths of 90°, 60° and 30°, respectively, together with the corresponding experimental points ($\bigcirc$, $\square$ and $\triangle$, respectively). Acquisition time, 1.2 s. Reproduced with permission of Academic Press

non-equilibrium pulse conditions. Because this is a unique time–pulse criterion, it is not useful in the general case where we have a number of nuclei with differing relaxation times. It is very evident that one should not consider using other than 90° pulses for doing any quantitative measurement where NOE is present and one can conclude from the simple benzene case that a t_3 delay ca 2.5 times the T_{1C} value is sufficient to achieve full magnetization. This is not true in general, however—in addition to the ^{13}C relaxation time, one must consider proton relaxation and cross relaxation between protons and carbon.

It is worthwhile considering other instrument and sample preparation variables before recommending an optimum procedure for small molecule characterization. Table 1 shows measured T_1 values for ethylbenzene under different conditions and at different field strengths. The effect of using oxygenated samples (air saturated at room temperature) compared with degassed samples is dramatic. The T_1 measured under NOE suppressed conditions (Table 1) should be a reliable estimate of the correct recycle rates for a quantitative experiment. The effect of proton–carbon cross-relaxation is seen in the T_1 measured under fully decoupled conditions (column 3). Since proton decoupling was commonly used to obtain many of the published values, one should be careful and add ca 20% to any T_1 estimates that are taken from the literature.

TABLE 1
^{13}C Relaxation times (seconds) for ethylbenzene

Relaxation studies were performed using an inversion–recovery sequence. Arrayed delay values were chosen between $0.5T_1$ and $2T_1$. The delay between pulse cycles was 500 s for the degassed samples and 250 s for the others.

Carbon	Degassed, NOE suppressed (90 MHz)	Oxygenated, NOE suppressed (90 MHz)	Oxygenated, decoupled (90 MHz)	Oxygenated, NOE suppressed (400 MHz)
1	83.38	54.40	52.76	49.5
2,6	18.57	15.48	13.14	15.2
3,5	16.63	15.24	13.84	15.6
4	14.40	13.47	9.10	11.4
CH$_2$	17.96	14.70	13.88	14.9
CH$_3$	11.91	7.59	8.52	9.1

Because quaternary aromatic carbons or other anisotropic carbons often determine the repetition rate of the experiment, it is desirable to use the highest field possible for quantitative measurements. If one has an estimate of T_1, a recycle rate of $6.8T_1$ is recommended for fast-relaxing carbons that have directly bonded protons[3]. For slow-relaxing quarternary carbons, $4.6T_1$ is adequate. In many cases, one does not have an estimate of T_1; for unknown cases, a recycle time of 300 s is a good compromise. In the case of ethylbenzene, an experiment consisting of four pulses with a 300 s recycle time gives a quantitative measurement of the carbons with a standard error of less than 0.5%. Figure 2 is useful for estimating relaxation times.

Signal-to-noise ratio is lost or gained as a function of $n^{1/2}$ (where n is the number of acquisitions), so choosing an incorrect recycle estimate can have specific consequencies. Thus, overestimating T_1 (i.e. acquiring more slowly than necessary) by a factor of 1.5, 2.0 or 3.0 causes a loss of signal-to-noise ratio by a factor of 1.2, 1.4 or 1.7, respectively, in a given time. The underestimation of T_1 (acquiring too rapidly) causes an error in the value determined for the relative integrated areas of resonances. It has been reported[4] that underestimating T_1 by factors of 1.5, 2.0 and 3.0 causes errors of approximately 6, 16, and 40%, respectively, in the integrated area of carbon resonances. Obviously, one must balance the consequences of making an estimate of T_1 against actually measuring the relaxation times of the carbons of the molecule or mixture of interest.

2.2 Addition of relaxation reagents

One of the most widely used methods for decreasing the relaxation time for performing quantitative measurements is provided by the addition of paramagnetic compounds. The effect of paramagnetic oxygen on T_1, as shown in Table 1,

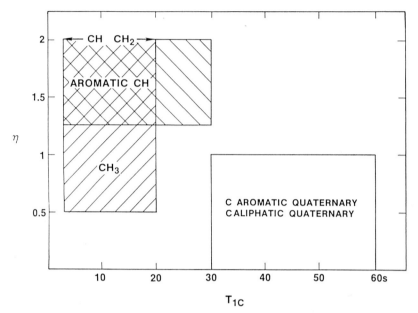

Fig. 2. Schematic graph of the possible nuclear parameters η and T_{1C} for hydrocarbons. Reproduced with permission of Academic Press

is clear. It was originally thought that the addition of stronger paramagnetic sources would both shorten T_1 and eliminate the NOE. Although this is theoretically possible, it was quickly realized that, realistically, the limit of solubility of paramagnetic compounds and the linewidth broadening caused by high concentrations of the reagents would make it impractical to use them at a sufficiently high level to eliminate NOE. Table 2 shows results published by Shoolery[1] on the effects of adding tris(acetoacetato)chromium [Cr(acac)$_3$] to acenaphthene. Comparison of the intensity results from a qualitative experiment with those done in the presence of 0.4 M Cr(acac)$_3$, under continuous decoupling conditions, shows that, although the large amount of the paramagnetic compound helps the measurement, the results are still not acceptable. On the other hand, the use of more modest amounts of Cr(acac)$_3$ together with gated decoupling gives a more quantitative spectrum than that acquired with a 400 s pulse delay and gated decoupling. Clearly, the use of the paramagnetic relaxation reagents is dramatic and effective.

Figure 3 shows the level of Cr(acac)$_3$ that should be used for the experiment. There is a rapid change in the ratio of aromatic to aliphatic carbon up to the point where 0.1 M Cr(acac)$_3$ is added. Higher concentrations of Cr(acac)$_3$ further improve the agreement between the aliphatic and aromatic intensities but the approach to unity appears to be asymptotic. Since the linewidth

TABLE 2
Integrated ^{13}C spectra of acenaphthene[1]

Ring position	Qualitative[a]	Qualitative[a] [0.4 M Cr(acac)$_3$]	Quantitative[b] (0.1 M Cr(acac)$_3$)	Quantitative[c]
1,2	2.00	2.00	2.00	2.00
3,8	1.64	2.00	1.98	2.00
4,7	1.62	1.96	1.96	1.98
5,6	1.57	1.96	1.94	2.02
9,10	0.51	1.87	1.94	2.00
11	0.19	0.94	0.96	0.96
12	0.19	0.94	0.97	0.90

[a] 20° pulse, 1 s repetition.
[b] 90° pulse, 2 s repetition, gated decoupler.
[c] 90° pulse, 401 s repetition, gated decoupler.

deteriorates at higher levels of Cr(acac)$_3$, 0.1 M is an excellent compromise. This level of paramagnetic species has been consistently adopted and recommended by a number of workers[5-10].

There is a substantial body of literature available on the use of paramagnetic relaxation agents. Paramagnetic reagents have been used most extensively for the analysis of complex mixtures of hydrocarbons[11-28]. However, there have been several articles that recommend caution in the use of paramagnetics[10]. In

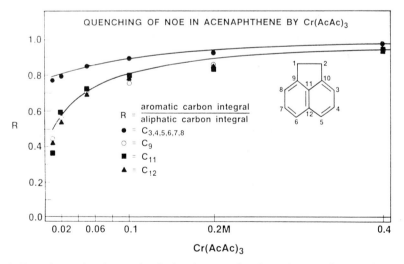

Fig. 3. Experimental points and calculated curves for the ratio (R) of aromatic carbon integrals to the aliphatic carbon integral as a function of Cr(acac)$_3$ concentration. Reproduced with permission of Pergamon Press

particular, care must be taken if the material of interest has a functional group that is likely to bond directly or displace partially the ligands present on the transition metal. Alcohols, acids and amine bases may have their relaxation times unusually affected by paramagnetic agents. Other paramagnetic reagents have also been suggested [particularly Gd(acac)$_3$ and Fe(acac)$_3$ as being more effective than Cr(acac)$_3$] because of their greater effect per unit concentration.

In summary, the addition of paramagnetic relaxation reagents is a highly successful and widely used quantitative NMR method. The normal concentration of paramagnetic reagent used is 0.1 M Cr(acac)$_3$. This reduces the relaxation time to less than 1 s for all types of carbon nuclei. The recycle delay period should, therefore, be at least 5 s for good quantitative values to be achieved. However, as the field strength and molecular weight increase (> 500), the relaxation times for quaternary carbons are already short. In these cases, the use of paramagnetic compounds could be detrimental. Levy has shown that for a molecule as low in molecular weight as cholesterol, there is no advantage in adding a paramagnetic compound[2,10].

2.3 ^{13}C NMR vs other analytical approaches

Perhaps the best test of quantitative ^{13}C NMR spectroscopy lies in its ability to provide accurate and precise analysis of mixtures. Tables 3 and 4 show results from two groups who compared quantification by ^{13}C NMR with other

TABLE 3
Composition (%) of linseed oil[1]

Method	Stearic	Oleic	Linolenic	Linolenic[a]
^{13}C NMR[a]	55.6	14.2	20.0	10.2
GC analysis[b]	55.3	15.3	19.4	9.9

[a] 0.04 M Cr(acac)$_3$, 90° pulse, 5 s, recycle delay, gated decoupling.
[b] GC analysis is the sum of C_{18} and C_{16} saturated acids while NMR gives the sum of all saturated acids.

TABLE 4
Composition (%) of mixed aromatic hydrocarbons[4]

Method	Tetralin	Phenanthrene	Hexahydropyrene
^{13}C NMR[a]	49.6	30.0	20.4
Gravimetric	50.5	30.1	19.4

[a] 0.11 M Cr(acac)$_3$, 90° pulse, 2 s recycle delay, gated decoupling.

analytical methods or with a known sample composition[1]. In both cases, they had previously determined the longest relaxation times of all the ^{13}C nuclei and acquired their spectra at five times the longest T_1 relaxation value. Similar results have been reported elsewhere.

2.4 Flow systems

An interesting mechanical solution to overcoming long relaxation times was demonstrated by Laude et al.[29], who showed that a flowing system, similar to that developed for LC liquid chromatography–NMR, can be used adequately for quantitative measurements. In considering a flowing system, the controlling factor, instead of the delay between pulses, is the amount of 'premagnetization time' before the sample can be examined. The sample must be in the field for five times the longest T_1 before the experimental measurement can be made. The sample can, of course, remain outside the field for as long as desirable to allow complete NOE quenching. Laude et al.'s device had a 22 μl measurement volume and a 10 ml premagnetization volume. The system required no major modification of their commercial instrument. A simple coil of Teflon tubing was used for the premagnetization area. This was connected to the bottom of a capillary tube microcell which was inserted into the convential NMR probe. The sample was then pumped with a peristaltic pump at a rate that allowed complete relaxation.

A comparison of static and flowing solution measurements is shown in Table 5 for acenaphthene and for ethyl benzoate. Clearly, the quantitative measurements are acceptable for both flowing and static conditions. However, the time reduction in acquiring spectra of equivalent signal-to-noise ratio using flowing conditions is truly impressive. On average, flowing spectra were acquired 30 times faster than static samples. The author cautions that, as relaxation times become shorter, the advantage of flow analysis decreases; however, it must always remain at least a factor of five better than static experiments. The major disadvantage of this method is clearly the total amount of sample that is required for the analysis. However, there are many instances where trading sample volume for analysis speed is acceptable.

2.5 Use of internal standards

Another accepted practice in quantitative measurement is the use of calibrated internal standards[30–33]. Internal standards are most widely used when all of the resonances for a sample cannot be observed or interpreted. This commonly occurs when the sample does not totally dissolve, as is often the case in drug and

TABLE 5
Comparison of static and flow quantification

	Acenaphthene			Ethyl benzoate	
Carbon No.	Static[a] (280 min)	Flow[b] (15.8 min)	Carbon No.	Static[c] (83.4 min)	Flow[d] (3.5 min)
			1 (quaternary)	99	101
1,2	203	204	2,6	201	199
3,8	200	206	3,5	202	205
4,7	198	208	4	99	98
5,6	202	202	7 (C=O)	100	97
9,10	200	198	8	99	100
11	96	98	9	99	101
12	96	101			

[a] 90° pulse, 300 s delay, NOE suppressed.
[b] 90° pulse, 8 s delay, NOE suppressed, 25 ml premagnetization, 0.5 ml min⁻¹ flow.
[c] 90° pulse, 125 s delay, NOE suppressed.
[d] 90° pulse, 0 s delay, NOE suppressed, 25 ml premagnetization, 12.5 ml min⁻¹ flow.

polymer analysis. Internal standards are also used when portions of the spectra are highly overlapped. If one or more resonances can be assigned, then they can be integrated against the internal standard. In the author's laboratory, this practice is used when qualitative identifications of complex mixtures have been made by gas chromatography–mass spectrometry and it is desired to quantitate the components by NMR. Another common reason for using internal standards is to avoid waiting for all of the carbon nuclei in a molecule or mixture to relax. Protonated carbons of higher molecular weight (>200) compounds commonly have relaxation times shorter than 1 s. As long as it is possible to find an internal standard with a short relaxation time, e.g. ethylene glycol, then the pulse recycle time on short relaxing carbons can be ignored. The weight percentage of any component can be calculated simply as follows:

$$\text{Wt-\%} = MW_{unk} \times \frac{I_{unk}}{I_{std}} \times \frac{Eq_{std}}{Eq_{unk}} \times \frac{wt_{std}}{MW_{std}} \times \frac{1}{wt_{unk}} \times 100$$

where MW_{unk} and MW_{std} are the molecular weights of the unknown and standard, Eq is the number of equivalent nuclei represented by the unknown and standard resonances measured and wt is the weight of the unknown or standard. The greatest limitation of this method is finding a good, non-interfering standard. Some that have been suggested are hexamethyldisiloxane, cyclohexane, adamantane, ethylene glycol, dioxane, dichloromethane, benzene and carbon disulfide.

2.6   Polymers and mixtures of stereoisomers

Quantitative analysis of polymers present no unusual challenges compared with previously mentioned methods. Specific results in this area have been treated by Randall[34,35]. In general, one does three types of quantitation for polymer analysis: composition, sequence distribution and tacticity distribution. Composition analysis is very similar to the analysis of mixtures of small molecules except that the relaxation times are usually much shorter. For typical vinyl polymers, the rate-limiting carbons are normally carbonyls, quaternary aromatics or nitriles, which normally have relaxation times of less than 3 s. Quaternary aliphatic carbons commonly have relaxation times of less than 1 s, and all protonated carbons have relaxation times of less than 1 s. Condensation polymers also normally have short relaxation times (less than 3 s) at high fields. At low fields, certain polyesters have relaxation times as long as 5 s for the carbonyl carbons. Protonated carbon resonances normally have relaxation times of less than 1 s. Therefore, if it is possible to use protonated carbons for the quantitation, recycle times of 5 s are acceptable. Almost universally, a recycle rate of less than 20 s will provide adequate quantitation. NOE factors do differ significantly from carbon to carbon; in particular, NOE must be suppressed if one is to compare the intensities of quaternary and protonated carbons.

Measurements of sequence distribution and particularly tacticity can be exceptions to the above rules[36,37]. The carbons from a specific functional group have different relaxation times, but the differences are so small that they cannot be accurately measured. In analysis for tacticity, where one is actually quantitating a mixture of different stereoisomers, the experiment can be run under conditions to maximize the signal-to-noise ratio (S/N). This is often helpful because, for these analyses, one often has to use a higher than normal digitization rate and narrow the observation window. Results for a tacticity analysis of polyacrylonitrile are presented in Table 6[38,39]. In this case, triad tacticity was measured and used to determine the probability of *meso* placement (given as calculated values). Calculated values are compared with observed values for tacticity distribution. The agreement between observed/calculated values and values obtained from ^{1}H runs for βd_2-polyacrylonitrile is excellent[38]. Therefore, in the special situations of stereoisomer analysis, the requirement for full relaxation and NOE suppression may be avoided.

With caution, the same assumption can be made for sequence distribution analysis. If one compares the relaxation times for the same functional group carbon from different sequence units, they normally do not vary by more than 10 %. Similarly, NOE factors are the same. Therefore, if one wishes to focus only on a narrow area of the spectrum, some instrument time can be saved by neglecting the requirement for full relaxation between acquisitions and suppression of NOE. However, since sequence analyses are usually combined with composition analysis, a non-equilibrium pulse condition is not commonly used.

TABLE 6
Tacticity distribution in polyacrylonitrile

Chemical shift (ppm)	Observed[38] ^{13}C	Calculated ($Pm = 0.527$)	Observed[39] ^{1}H
	Triad		
118.2	mm 0.278	0.278	0.27
117.9	mr 0.495	0.499	0.50
117.6	mr 0.226	0.224	0.23
	Pentad		
118.3	mmmm 0.076	0.077	0.07
115.2	mmmr 0.130	0.158	0.14
118.1	rmmr 0.070	0.062	0.06
118.0	mmrm 0.137	0.183	
117.9	rmrm 0.241	0.248	0.50
117.8	mmrr 0.117	0.111	
117.7	mrrm 0.060	0.062	
117.6	mrrr 0.110	0.111	0.23
117.5	rrrr 0.056	0.050	

3 ^{13}C NMR IN SOLIDS

3.1   Quantification of CP–MAS spectra

Quantitation in solids presents some additional challenges not common to liquid samples. Firstly, there is the question of multiple domains in the solid. In solution, substantial differences in proton content cause differences in relaxation times within a molecule, but in solids we have to be concerned with both directly bonded protons and protons adjacent in space. We can expect differences in relaxation times even for compounds which are simply different isomers. There are even differences in relaxation times for the same solids having different crystallographic forms.

The solid-state experiments most commonly performed are those in which magic angle spinning (MAS) and cross-polarization (CP) are used. MAS is used to remove ^{13}C chemical shift anisotropy and CP is used for signal-to-noise enhancement. MAS alone has no effect on quantitation, except for creating problems with spinning sidebands; CP, on the other hand, can have substantial effects on signal intensities. In certain cases, MAS suppresses the CP effect.

In the CP experiment, the proton spins are spin-locked by applying a 90° pulse and then applying a locking pulse in the X-Y plane (90° phase shifted from the first pulse) of pulse duration τ. While the proton spins are locked, the carbon spins are simultaneously excited with a pulse of duration τ. This fulfils the Hartmann–Hahn condition of $B_{1C\gamma_C} = B_{1H\gamma_H}$ (γ_C and γ_H are the magnetogyric ratios and B_{1C} and B_{1H} are the magnitudes of the spin-locking fields). The effect of Hartmann–Hahn matching is to produce an enhancement of the carbon magnetization. The time period (τ) during which the proton spins and carbon spins are in contact is one of the key variables which determines whether the ^{13}C spectra acquired are quantitative. Another key variable is the time between repeat applications of the spin-lock cycle. In certain cases, the actual length of the exciting B_1 pulses can also affect quantitative results[40]. The contact time governs intensity for a homogeneous material and both contact time and repetition rate govern intensity in heterogeneous materials.

The cross-polarization technique affects carbon signal intensity in two ways. Firstly, there is a direct signal enhancement of a factor of up to 4 for ^{13}C. Secondly, there is a signal enhancement because the repeat cycle of the experiment is determined by the proton relaxation times. Proton relaxation times are generally much faster than carbon relaxation times and, therefore, S/N can be accumulated at a more rapid recycle rate than if ^{13}C was observed directly without CP.

For quantitative measurements, the primary variables are the proton relaxation rate in the solid (normally referred to as T_{1H}) and the rate of spin transfer from protons to carbon (T_{CH}). As in the liquid, the experiment must be repeated no faster than five times T_{1H}. It can also be assumed that, in a given solid, all proton relaxation times are equal owing to dipolar-spin exchange. In the liquid, spin polarization of carbon is direct and rapid. In the cross-polarization experiment, spin polarization depends on the ^{1}H–^{13}C distance and typically takes 0.1–1 ms. Competing with spin polarization build-up are relaxation processes of both proton and carbon in the rotating frame $T_{1\rho}$. To achieve a quantitative spectrum, we must establish the condition where $T_{CH} < t_{C\rho} < T_{1\rho H}$, $T_{1\rho C}$ (where $t_{C\rho}$ = contact time or cross-polarization time).

The rate of cross-polarization buildup for a ^{13}C nucleus is proportional to both the number and distance of proton spins. It also depends on the internal motion of the various groups within a molecule. In general, the rapid rotation of methyl groups within a solid reduces the rate of polarization transfer to approximately that a methine carbon. Qualitatively, we would expect that T_{CH} would fall in the sequence $CH_{3(static)} > CH_2 > CH > CH_{3(rotating)} > C_{(non-protonated)}$. As in liquids, therefore, the quantitative experiment is dominated by a non-protonated carbon.

Several workers have published results on CP–MAS studies on simple solids[41,42]. The most systematic study is contained in two companion papers by Alemany et al.,[41,42] where a series of solids were considered. They studied the effect of varying contact time on intensity. Their results for (4-ethoxyphenyl)-acetic acid (Table 7) are representative of those presented in their papers.

TABLE 7
Relative signal intensities in (4-ethoxyphenyl)acetic acid[41] as a function of contact time

Contact time (ms)	Relative intensity				
	C—O	C—4	C—2,6	C—1	C—3,5
	Theoretical				
	10	10	20	10	20
	Experimental				
3.000	10.8	9.8	16.9	10.2	20.0
2.250	10.4	10.2	18.5	10.1	20.2
1.500	10.8	10.9	18.3	10.8	19.1
1.000	10.8	10.1	18.6	9.8	20.1
0.700	10.1	9.7	19.1	10.2	19.8
0.500	9.6	8.4	20.0	9.0	20.2
0.400	8.5	7.9	20.2	9.4	20.6
0.300	8.2	6.9	20.3	8.6	21.0
0.225	6.8	5.6	21.4	6.8	23.0
0.150	5.1	4.4	22.6	6.7	24.1

Contact time (ms)	OCH_2	CH_2CO	CH_3	Std error	Rel. std error
	Theoretical				
	10	10	10		
	Experimental				
3.000	9.4	10.4	12.5	1.5	11.1
2.250	8.7	10.3	11.6	0.9	8.0
1.500	9.2	9.8	11.0	1.0	7.7
1.000	10.0	9.2	11.5	0.8	7.1
0.700	10.1	9.7	11.2	0.6	4.9
0.500	10.4	10.0	12.3	1.1	10.7
0.400	10.4	10.7	12.4	1.3	13.0
0.300	11.1	10.8	13.2	1.9	18.4
0.225	11.9	10.9	13.7	2.9	27.5
0.150	12.2	12.7	12.2	3.7	33.4

Comparing relative intensities between the C-4 carbon and the ethoxy carbon (Figure 4) shows that any contact time between 0.5 and 2.25 ms would give acceptable relative intensities for structure proof.

Several important points are apparent from these data: (i) the ethoxy carbon ($T_{CH} = 0.11$ ms) achieves full polarization very rapidly and begins to lose intensity at longer contact times owing to $T_{1\rho}$ effects; (ii) the C-4 carbon ($T_{CH} = 0.44$ ms) probably does not achieve full intensity before $T_{1\rho}$ becomes dominant. Therefore, although not true in this instance, it is possible that there will be cases where one cannot achieve quantitative results because protonated carbons with short $T_{1\rho}$ values lose too much intensity at the longer contact times

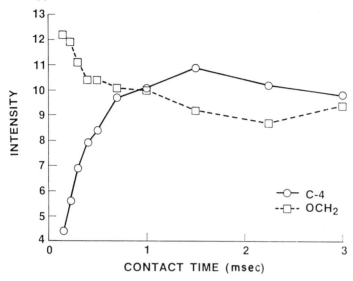

Fig. 4. Comparison of intensity of ring carbon C-4 and OCH$_2$ in (4-ethoxyphenyl)acetic acid versus contact time

required to give non-protonated carbons full intensity. It has been shown that when carbons are four or more bonds away from a protonated site, correct relative intensities *cannot* be obtained[42].

For (4-ethoxyphenyl)acetic acid, Alemany *et al.* did not study extensively the effects of T_{1H} on the spectra. However, they stated that 1,8-dimethylnaphthalene and 2,6-dimethylnaphthalene have markedly different proton relaxation values. Under identical recycle conditions (3 s delay), the S/N of spectra of 2,6-dimethylnaphthalene is much lower than that of 1,8-dimethylnaphthalene. For structure proof, both compounds would give correct intensities using a contact time of 2.25 ms and a recycle time of 3 s. However, quantitation of a mixture of these compounds would require a recycle delay in excess of 60 s. They recommended that simple molecules can be studied adequately with the use of recycle delays of 3 s and contact times of 2.25 ms.

3.2   Coal and related materials

One of the most widely used quantitations using CP–MAS is in the area of coal and coal-related materials such as peats, lignins and cellulose[48–54] (for a more detailed discussion of the analysis of fossil fuels, see Chapter 4). In particular, the aromaticity (percentage of aromatic carbons) is normally examined by NMR spectroscopy and, for coal, there are several potential problems for quantitative measurements. Firstly, coals are not homogeneous materials and there will not be complete spin diffusion throughout the solid. Another potential problem is the fact that a significant level of paramagnetic species (ca 1×10^{17} spin g^{-1}) is present in coal and these can cause rapid relaxation of both the ^{1}H and ^{13}C nuclei in their vicinity.

Several workers have examined the coal problem in detail. Early work by Maciel and Sullivan[48] on US coal showed that CP–MAS accurately measured coal aromaticities. Aromaticities were determined using CP–MAS and also by direct ^{13}C excitation ($5T_1$ waiting between pulses and gated decoupling). The values were found to be the same, within experimental error. In subsequent similar studies Dudley and Fyfe[49] found aromaticities higher (0.07–0.16) when they were measured using traditional methods. Packer et al.[50] argued that the primary problem in coal analysis using CP–MAS is the occurrence of components with very short $T_{1\rho}$. Correct quantitation would require $T_{1\rho}$ values greater than 1 ms for all protons. Direct measurement of ^{1}H $T_{1\rho}$ values on coal show that 20–50% of the protons have $T_{1\rho}$ values less than 1 ms; however, the reason for the very short $T_{1\rho}$ values is not entirely clear. Several studies have shown that the addition of a paramagnetic compound can significantly reduce $T_{1\rho}$. Additionally, paramagnetic compounds also affect the T_2 of the protons. Newman has shown that if the T_2^* of protons can vary substantially in the different solids and if the B_1 field is not sufficiently strong, then significant relaxation can take place during the excitation pulses. In areas of high paramagnetic species, $T_{2(1H)}^*$ may be short enough to inhibit quantitative excitation.

Studies on coal show that one must be extremely cautious in using CP–MAS for performing quantitative measurements. When paramagnetic species are present and/or the material is heterogeneous, it is prudent to check (or calibrate) the CP–MAS analysis using traditional full relaxation studies. However, full relaxation studies can require times up to two orders of magnitude longer to acquire data of similar quality in solids and, therefore, should be used sparingly.

3.3   Solids with long T_1

The difficulty with full relaxation studies can be appreciated if we consider a solid of very low proton density. Materials such as diamond and silicon carbide represent the ultimate solid materials with long T_1 relaxation times[55,56]. A progressive saturation T_1 study for ^{29}Si and ^{13}C of silicon carbide is shown in Figure 5. Two dramatic effects are observed. Firstly, there appears to be a rapidly relaxing material with a T_1 of ca 1s. Secondly, the bulk silicon and carbon T_1 values are in excess of 2000 s! Therefore, in this case, a recycle rate of 3 h is needed to achieve quantitative measurements by ^{13}C NMR. Although these spectra show usable S/N after only 16 pulses, it still requires 48 h for the data to be acceptable for true quantification. Such very long relaxation times are not unusual for spin-$\frac{1}{2}$ nuclei in non-proton-containing solids.

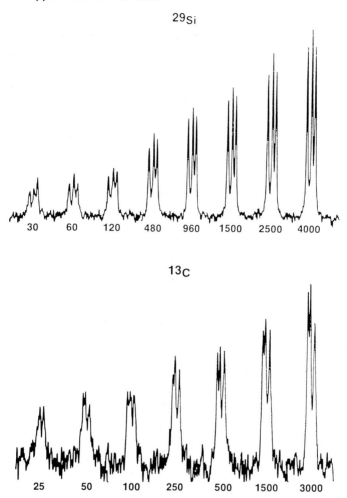

Fig. 5. ^{29}Si and ^{13}C NMR spectra of silicon carbide as a function the delay (seconds) following saturation

Quantitative measurements in solids have been done most routinely on polymers, which present an almost ideal system for examination by NMR. Polymers typically contain a very high proton density, are commonly homogeneous and usually have a very low paramagnetic content. The CP–MAS experiment was pioneered by Shaefer and co-workers[57,61] and they and others have achieved almost universal success in effecting reliable ^{13}C quantitation for most types of carbon in the polymer. No unusual precautions are necessary for quantitation of solid polymers and, typically, contact times between 0.5 and 1 ms give acceptable results. The most difficult cases are those in which significant portions of the polymer have low glass transition temperatures. The high mobility of these segments coupled with high-speed spinning makes it impossible to establish a Hartmann–Hahn matching condition. A quick method of checking for this problem is to examine the MAS proton spectrum. Any portions of the polymer giving a resolved ^{1}H spectrum will normally not be observable in the CP–MAS spectrum. This is most often found for rubber-reinforced polymers, but there are cases where only one carbon (often a rapidly rotating methyl) will not cross-polarize. Applications of the CP–MAS experiment to polymers have been well reviewed[62].

4 PRACTICAL CONSIDERATIONS

4.1 Pulse width

Having established relaxation criteria for good quantitation, there remain certain spectrometer and data handling conditions that can limit the quality of quantitative measurements. The first criterion is that the pulse power be strong enough to deliver linear excitation across the full spectrum of interest. The excitation pulse length, normally 90°, must be less than the inverse of the spectral bandwidth. The desirable pulse width for ^{13}C spectra would be less than 25 μs. If the available power–probe combination does not allow such a short 90° pulse, then it will be necessary to use a smaller pulse angle, which will cause loss of S/N (see Figure 1).

4.2 Filter bandwidth

The filter bandwidth setting is a cause of non-linearity in intensity between one end of a spectrum and the other. This is the most common problem on modern spectrometers. Vendors recommend (and in many cases automatically set) filter settings which are optimized for maximum S/N, i.e. designed to reject noise from outside the spectral window. Figure 6 shows a series of proton spectra of the ASTM reference sample (see Section 1) taken with the spectrometer carrier frequency offset successively by 1 ppm. In Figure 6A, the manufacturer's recommended filter settings were used. Figure 6B shows the results obtained using a filter bandwidth equal to twice the spectral bandwidth. Clearly, operating under the latter conditions gives a reproducible signal intensity across the entire spectral range whereas a reduced filter bandwidth attenuates signals close to the edges of the window. This test is quickly performed for any nucleus. For ^{13}C spectra, either the ASTM sample or Cr(acac)$_3$-doped benzene is a good

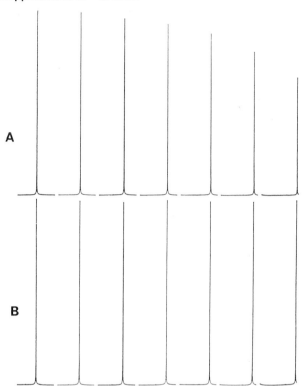

Fig. 6. ^{1}H NMR spectra of dioxane as a function of offset from the carrier frequency. The transmitter offset increases by 400 Hz between successive experiments. (A) Vendor-recommended filter setting; (B) filter setting equal to twice the sweepwidth. The sample run was the ASTM reference samples (see Section 1). The transmitter offset was arrayed at 400 Hz intervals. Pulse, 500 μs; 20-s delay; 32 scans; absolute intensity

test sample. About ten points across the spectral region of interest should be tested. Problems of filter bandwidth and weak pulses can be easily observed using the test and it should be performed regularly on all spectrometers. It is also possible to calibrate for this loss in signal intensity[62].

4.3 Phase anomalies

Several instrument variables affect the ability to phase spectra. Filters, in general, will cause phase anomalies. In particular, older spectrometers had notch filters and crystal filters that would not allow pure absorption spectra to be produced. The filter-limiting bandpass, if not set as above, will cause phase distortions. The delay between the pulse and the beginning of acquisition will also affect phase settings. Normally, this can be corrected if the delay is set to approximately one dwell time, and this is often done automatically by software.

Phase distortion can also be caused by pulsing too rapidly[64]. Freeman and Hill[65] have shown that if the interval between repeated pulses is less than the spin–spin relaxation times, phase anomalies can appear in the spectrum. This could occur in cases where the pulse recycle rate was set for rapidly relaxing nuclei with the intention of disregarding signals from slowly relaxing nuclei. The condition can be partly alleviated by applying line broadening prior to FT; however, the best solution is to use longer delay times. Another cause of phase distortion of narrow lines arises from under-sampling of the free induction decay[66]. If the spectra are adequately sampled for good integration (see Section 4.6), then phase distortion should not occur. However, this phase distortion could be a symptom that it is necessary to increase the number of acquired data points by a factor of two. It is wise to ensure that correct phasing can be done over the complete spectrum. A test sample consisting of CS_2, $CDCl_3$ and hexamethyldisiloxane with 0.1 M $Cr(acac)_3$ will ensure adequately that correct phasing can be done over the entire spectral range. Similar mixtures have been proposed by others[62].

4.4   Baseline artefacts

A consistent problem, particularly in earlier generation spectrometers, is the lack of baseline flatness. Normally, two non-related phenomena cause baseline humps; one is instrumental and the other is due to sample inhomogeneity. Acoustic ringing in the probe causes a broad hump under the whole spectrum. This can normally be alleviated either by increasing the delay between the pulse and the start of data acquisition or by using a trapezoidal weighting function to suppress points at the beginning of the FID. The second major cause of baseline humps is solids (often not visible) in the sample. Filtering the sample through glass-wool in a pipet will remove most solids and is recommended for samples prepared with added $Cr(acac)_3$. Nevertheless, persistent artifacts may still be observed in spectra from early generation spectrometers and these can be corrected by substracting a fourth-order polynominal, first suggested by Pearson[67]. The correction procedure works very well when one is removing broad humps from an area of sharp lines; however, it is less successful if there are real, broad features in a spectrum, as with polymers.

4.5   Signal-to-noise ratio and digital resolution

Two experimentally controlled parameters that strongly affect integral precision are the signal-to-noise ratio (S/N) and digitization level in the spectrum. Sotak et al.[2] published an excellent study that addressed these questions. They showed that precision does not improve after achieving an S/N (r.m.s.) of greater than ca. 35:1. However, the precision drops very rapidly as the S/N drops below this level.

Digital resolution is governed by the spectral bandwidth and the size of the data set. In general, spectra should always be zero filled to maintain the number of real points (in the transformed spectra) equal to the number of acquired points. The sampling interval (hertz per point) should be one-third of the full width at half-height for the narrowest peak of interest in the spectrum. Another way of stating this is that there should be at least two data points above the half-height of any peak. Additional sampling points do not aid the integration and, in fact, will degrade it because the S/N will decrease owing to the sampling of more noise from the free induction decay. For most carbon spectra, a sampling interval of 1 Hz per point is appropriate.

4.6 Integration

There are a number of software considerations in performing good quantitative measurements. The ease and quality of phase correction in the spectrum are key factors. Automatic phase corrections are now appearing in some software packages and, although these work well for sharp lines, in general the bias of an integral is a much better judgment of the correctness of phase than judging the spectrum for pure absorbance by eye. It may be necessary to rephase and integrate the spectrum over narrow regions, depending on the available software. The integration packages supplied by most vendors use a running sum of the points above zero. This software works adequately if peaks are well separated and can be well phased. A better estimate is obtained using a Simpson's rule algorithm.

In cases where peaks are not fully resolved, some deconvolution or curve-fitting procedure must be used. If the S/N is good, deconvolution (resolution enhancement) is most appropriate and requires the fewest assumptions. More often, curve fitting, using a mixture of Gaussian or Lorentzian lineshapes, is used[2,69]. Curve fitting is highly successful and gives good precision, as long as less than one-third of the peak areas are overlapped. Curve fitting is now commonly available as part of commercial software packages.

5 ACKNOWLEDGEMENTS

I acknowledge the helpful discussions with my co-workers, T. Hammond, S. Mocadlo and K. Benton, in the preparation of this chapter. I am also indebted to R. Boyer, T. Hammond and S. Mocadlo for the data in this chapter taken at Standard Oil.

REFERENCES

1. J. N. Schoolery, *Prog. Nucl. Magn. Reson. Spectrosc.*, **11**, Pt 2, 79 (1977)
2. C. H. Sotak, C. L. Dumoulin and G. C. Levy, *Top. Carbon-13 NMR Spectrosc.*, **4**, 91 (1984).
3. S. Gillet and J. J. Delpuech, *J. Magn. Reson.*, **38**, 433 (1980).
4. D. J. Cookson and B. E. Smith, *J. Magn. Reson.*, **57**, 355 (1984).
5. T. D. Alger, M. Solum, D. M. Grant, G. D. Silcox and R. J. Pugmire, *Anal. Chem.*, **53**, 2299 (1981)
6. T. D. Alger, R. J. Pugmire, W. D. Hamill, Jr, and D. M. Grant, *Am. Chem. Soc. Div. Fuel Chem., Prepr.*, **24**, (1979).
7. J. J. Pesek, *Anal. Chem.*, **50**, 787 (1978).
8. F. El-Shahed, K. Doerffel and R. Radeglia, *J. Prakt. Chem.*, **321**, 859 (1979).
9. S. A. Sojka and R. A. Wolfe, *Appl. Spectrosc.*, **34**, 90 (1980)
10. G. C. Levy and U. Edlund, *J. Am. Chem. Soc.*, **97**, 4482 (1975).
11. M. Bouquet and A. Bailleul, *Fuel*, **65**, 1240 (1986)
12. A. H. A. K. Mohammed, J. M. A. Al-Rawi and K. Hankish, *Fuel Sci. Technol. Int.*, **4**, 345 (1986).
13. D. F. Kushnarev, V. M. Polonov, V. I. Donskikh, E. F. Rokhina and G. A. Kalabin, *Khim. Tverd. Topl.*, (2), 31 (1986)
14. M. B. Simirnov and B. A. Smirnov, *Neftekhimiya*, **25**, 402 (1985).
15. J. C. Boubel, T. Carillon, J. M. Daubenfeld, J. J. Delpuech and B. Neff, *Collect. Colloq. Semin. Inst. Fr. Pet. (Caract. Huiles Lourdes Resid. Pet)*, **40**, 345 (1984).
16. J. J. Delpuech, *NATO ASI Ser., Ser. C. (Magn. Reson.)*, **124**, 351 (1984).
17. S. P. Srivastava, A. K. Bhatnagar and G. C. Joshi, *J. Chem. Technol. Biotechnol., Chem. Technol.*, **33A**, 361 (1983).
18. D. J. Cookson and B. E. Smith, *Fuel*, **62**, 986 (1983).
19. D. M. Barnhart and D. A. Netzel, *Am. Chem. Soc. Div. Fuel Chem., Prepr.*, **27**, 233 (1982).
20. S. P. Srivastava, *J. Chem. Technol. Biotechnol.*, **32**, 614 (1982).
21. K. Saito, R. J. Baltisberger, V. I. Stenberge and N. F. Woolsey, *Fuel*, **60**, 1039 (1981).
22. S. Gillet, P. Rubini, J. J. Delpuech, J. C. Escalier and P. Valentin *Fuel*, **60**, 226 (1981).
23. S. Gillet, P. Rubini, J. J. Delpuech, J. C. Escalier and P. Valentin, *Fuel*, **60**, 221 (1981).
24. J. T. Joseph and J. L. Wong, *Am. Chem. Soc. Div. Fuel Chem., Prepr.*, **24**, 317 (1979).
25. J. T. Joseph and J. L. Wong, *Fuel*, **59**, 777 (1988).
26. Y. Maekawa, T. Yoshida and Y. Yoshida, *Fuel*, **58**, 864 (1979).
27. W. R. Ladner and C. E. Snape, *Fuel*, **57**, 658 (1978).
28. Y. Maekawa, T. Yoshida, Y. Yoshida and M. Imanari, *Nenryo Kyokai-Shi*, **56**, 351 (1977).
29. D. A. Laude, Jr, R. W. K. Lee and C. L. Wilkins, *Anal. Chem.*, **57**, 1286 (1985).
30. S. T. Eberhart, A. Hatzis and R. Rothchild, *J. Pharm. Biomed. Anal.*, **4**, 147 (1986).
31. J. Tamate and J. H. Bradbury, *J. Sci. Food Agric.*, **36**, 1291 (1985).
32. H. C. Chiang, P. L. Wang and K. F. Huang, *J. Chin. Chem. Soc. (Taipei)*, **30**, 117 (1983).
33. H.-C. Chiang and L.-J. Lin, *Org. Magn. Reson.*, **12**, 260 (1979).
34. J. C. Randall and E. T. Hsieh, *ACS Symp. Ser.*, No. 247, 131 (1984).
35. J. C. Randall, *Polymer Sequence Determination*, Academic Press, New York, 1977.
36. F. C. Schilling, F. A. Bovey, K. Anandadumarnt and A. E. Woodward, *Macromolecules*, **18**, 2688 (1985).
37. P. Sozzani and C. Olivia, *J. Magn. Reson.*, **63**, 115 (1985).

38. H. S. Fochler, J. R. Mooney, L. E. Ball, R. D. Boyer and J. G. Grasselli, *Spectrochim. Acta, Part B*, **41**, 271 (1985).
39. O. Matsuzaki and G. Matsuzaki, *J. Polym. Sci., Part A-1*, **10**, 826 (1972).
40. R. H. Newman, *J. Magn. Reson.*, **72**, 337 (1987).
41. L. B. Alemany, D. M. Grant, J. Pugmire, T. P. Alger and K. W. Zilm, *J. Am. Chem. Soc.*, **105**, 2133 (1983).
42. L. B. Alemany, D. M. Grant, R. J. Pugmire, T. P. Alger and K. W. Zilm, *J. Am. Chem. Soc.*, **105**, 2142 (1983).
43. C. S. Yanonni, H. P. Reiseman and G. Mariner, *J. Am. Chem. Soc.*, **105**, 6181 (1983).
44. K. W. Zilm and D. M. Grant, *J. Am. Chem. Soc.*, **103**, 2913 (1981).
45. S. J. Opella and H. M. Frey, *J. Magn. Reson.*, **66**, 144 (1986).
46. G. S. Harbison, P. P. J. Mulder, H. Pardoen, J. Lugtenburg, J. Herzfeld and R. G. Griffin, *J. Am. Chem. Soc.*, **107**, 4809 (1985).
47. J. R. Scheffler, Y. F. Wong, A. O. Patil, P. Y. Curtin and I. C. Paul, *J. Am. Chem. Soc.*, **107**, 4898 (1985).
48. M. J. Sullivan and G. E. Maciel, *Anal. Chem.*, **54**, 1606 (1982).
49. R. L. Dudley and C. A. Fyfe, *Fuel*, **61**, 651 (1982).
50. K. J. Packer, R. K. Harris, A. M. Kenwright and C. E. Snape, *Fuel*, **62**, 999 (1983).
51. J. M. Dereppe, *NATO ASI Ser., Ser. C (Magn. Reson.)*, **124**, 535 (1984).
52. C. M. Preston and B. A. Blackwell, *Soil Sci.*, **139**, 88 (1985).
53. E. W. Hagman and R. R. Chambers, *Am. Chem. Soc. Div. Fuel Chem., Prepr.*, **30**, 188 (1985).
54. T. Yoshida, Y. Maekawa and T. Fujito, *Anal. Chem.*, **55**, 388 (1983).
55. P. M. Henricks, M. T. Cofield, R. H. Young and J. M. Hewitt, *J. Magn. Reson.*, **58**, 85 (1984).
56. J. S. Hartman, G. R. Finley, M. F. Richardson and B. L. Williams, *J. Chem. Soc., Chem. Commun.*, **159** (1985).
57. J. Schaefer, S. H. Chin and S. I. Weissman, *Macromolecules*, **5**, 798 (1972).
58. J. Schaefer, E. O. Stejskal and R. Buchdahl, *Macromolecules*, **8**, 291 (1975).
59. J. Schaefer and E. O. Stejskal, *J. Am. Chem. Soc.*, **98**, 1031 (1976).
60. J. Schaefer, E. O. Stejskal and R. Buchdahl, *Macromolecules*, **10**, 384 (1977).
61. J. Schaefer and E. O. Stejskal. In G. C. Levy, (ed.) *Topics in Carbon-13 NMR Spectroscopy*, Vol. 3, Wiley-Interscience, New York, 1979.
62. C. A. Fyfe, *Solid State NMR for Chemists*, C.F.C. Press, Guelph, Canada.
63. J. M. Daubenfeld, J. C. Boubel, J. J. Delpuech, B. Neff and J. C. Escalier, *J. Magn. Reson.*, **63**, 195 (1985).
64. F. A. L. Anet, L. Kaozhen and T. Zhihong, *Magn. Reson. Chem.*, **25**, 439 (1987).
65. R. Freeman and H. D. W. Hill, *J. Magn. Reson.*, **4**, 366 (1971).
66. M. B. Comisarow, *J. Magn. Reson.*, **58**, 209 (1984).
67. G. A. Pearson, *J. Magn. Reson.*, **27**, 265 (1977).
68. J. K. Kaupinen, D. J. Moffatt, H. H. Mantsch and D. G. Cameron, *Appl. Spectrosc.*, **35**, 271 (1981).
69. T. Nakayama and Y. Fujiwara, *Anal. Chem.*, **54**, 25 (1982).

Analytical NMR
Edited by L. D. Field and S. Sternhell
© 1989 John Wiley & Sons Ltd

Chapter 4

Analysis of Fossil Fuels

C. E. Snape

Department of Pure and Applied Chemistry, University of Strathclyde, Glasgow, UK

1   INTRODUCTION

1.1   Relevance

The need to evaluate the chemical nature of petroleums, coals, oil shales and tar sands when considering their utilization, specification and formation is self-evident. The production of synthetic fuels from coals, oil shales and tar sand bitumens received much attention following the uncertainty over the future of petroleum supplies during the early 1970s. Fossil fuels are inherently complex and contain many aromatic, aliphatic and heteroatomic moieties. Moreover, their compositions vary markedly (see Table 1 for coals); petroleums generally contain much more hydrogen and less oxygen than coals[1], which are largely insoluble in common organic solvents[2]. References 1 and 2 provide excellent accounts of the origins, formation and classification of petroleums and coals.

Many of the constituents in oils, such as n-alkanes in petroleums, can be readily identified and measured using gas chromatography (GC) and mass spectrometry (MS), although prior fractionation by liquid chromatography (LC) is desirable in order to limit the number of structural possibilities[3]. However, the involatility and complexity of heavy oils (boiling above ca 450 °C) and solid fuels has meant that bulk compositional properties, such as the

TABLE 1
Typical elemental compositions of some fossil fuels

Fuel	Concentration (%, ash-free basis)				
	C	H	O	N	S
Petroleum residue	83	10	1	0.5	5.5
Tar sand bitumen	84	10	<1	0.5	5
Lignite	68	5.5	25	1	0.5
Sub-bituminous coal	74	5	19	1	1
High-volatile bituminous coal	82	5.5	9	1.5	1
Low-volatile bituminous coal	89	5	4	1	1
Anthracite	94	3	2	1	0.5

proportions of aromatic, naphthenic and paraffinic carbon, have had to be used for molecular characterization. Before the advent of NMR, bulk compositions were deduced from correlations with physical properties, such as density and refractive index, together with molecular weight[4] (n–D–M analysis). The present widespread use of NMR in fuel characterization arises from the fact that it is inherently quantitative, with high-resolution ^{1}H and ^{13}C NMR providing direct information on the concentrations of aromatic and aliphatic groups.

1.2 Scope

This chapter provides a relatively broad overview of the use of NMR in fuel analysis with emphasis on the impact of multi-pulse and solid-state techniques and on the quantitative reliability of results. The subject is too vast for a comprehensive coverage in the space available. Indeed, recent texts devoted solely to solid-state NMR of fuels[5] and NMR analysis of liquid fuels[6] have recently been published and space only allows a strictly non-mathematical description to be given here for the many advanced methods applied to fuels.

The scope has been limited to nuclei used for probing organic matter in fuels, although it is realized that ^{29}Si, ^{27}Al and ^{23}Na can be used to characterize minerals associated with coals and oil shales. The discussion focuses mainly on high-resolution techniques (Sections 2 and 3) but important applications of broadline ^{1}H NMR are also included (Section 3.3). The use of structural analysis schemes in which NMR data are combined with elemental compositions, molecular weights and other analytical results to derive average compositional data, such as the number of peripheral and bridgehead aromatic carbons, for soluble products is assessed critically (Section 4).

1.3 Brief history

Before Fourier transform (FT) spectrometers were commercially available, high-resolution applications were limited to ^{1}H measurements on oils and soluble products. Some reports date back to before 1960[7,8], although continuous sweep ^{13}C spectra were obtained as early as 1966[9]. As in other fields, the use of NMR in fuel analysis has expanded considerably during the past decade with the introduction of improved instrumentation, enabling a wide range of pulsed FT methods to be applied to both solid and liquid fuels. Two-dimensional and ^{13}C editing methods have enabled unambiguous peak assignments to be made for oils (Section 2). Not surprisingly, the advent of high-resolution solid-state ^{13}C NMR has led to numerous investigations on coals[10,11], oil shales and related insoluble materials, such as chars (Section 3). However, quantification has proved to be much less straightforward than in the case of solution-state ^{13}C NMR. Multinuclear methods have found only limited application for the direct determination of O, N and S environments in fuels, but prior derivatization has enabled concentrations of OH, NH and SH groups to be determined (Section 2).

Broadline ^{1}H NMR has found some important applications in fuel chemistry. It was recognized that the hydrogen contents of liquid fuels could be determined more accurately than by combustion methods and this led to the publication of the first ASTM method on NMR in 1978[12]. More recently, pulsed FT methods have been used to differentiate hydrogen in mobile and rigid substituents in solid coals[2,10] and high-temperature measurements have enabled the pyrolytic behaviour of coals and oil shales[10,13] to be investigated (Section 3.3).

2 SOLUTION-STATE MEASUREMENTS

2.1 Sample fractionation and preparation

As for GC-MS analysis[3], some prior fractionation is desirable in order to reduce structural complexity for the detailed NMR characterization of heavy oils and more intractable materials, such as bitumens and tars. Figure 1 summarizes the steps that have been used in numerous studies. For bitumens and primary coal conversion products (tars, extracts), the first step is usually solvent fractionation with an aromatic (benzene or toluene) and an alkane solvent (n-pentane or n-heptane), asphaltenes being defined as material soluble in benzene or toluene but insoluble in the alkane solvent used. Oils or material soluble in alkane solvents can be separated by ion-exchange chromatography into acidic and basic fractions and by adsorption chromatography into paraffins and various aromatic fractions.

In many respects, chloroform-d is by far the most convenient solvent for ^{1}H and ^{13}C experiments (Sections 2.2 and 2.3). It is therefore advantageous to improve the chloroform solubility of relatively intractable material, such as certain coal extracts, pitches and liquefaction products. After silylation and methylation, highly phenolic benzene- and toluene-insoluble material in primary coal liquefaction products has been found to be virtually completely soluble in chloroform[14]. For petroleum and coal tar pitches, which are much less phenolic but contain considerable concentrations of quinoline insolubles, solubility in chloroform for ^{1}H and ^{13}C analysis has been achieved via chlorination with a mixture of sulphuryl chloride (SO_2Cl_2) and sulphur monochloride (S_2Cl_2)[15,16].

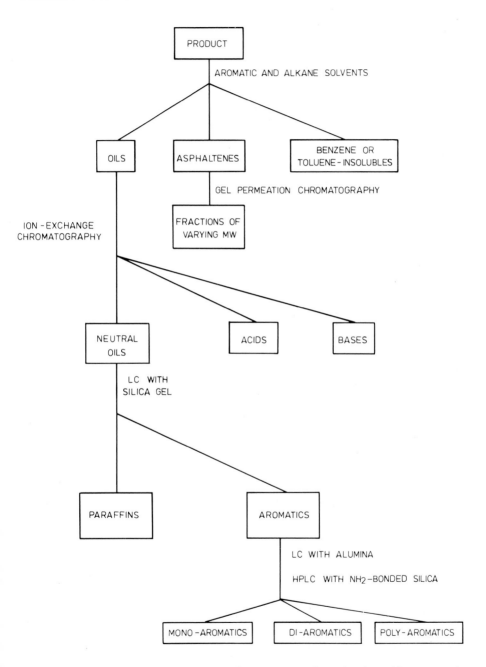

Fig. 1. Typical fractionation scheme used for heavy petroleum fractions, bitumens and coal-derived liquids

2.2 ^{1}H Measurements

2.2.1 General

FT spectra can be obtained using milligram quantities of sample but considerably more is required for CW spectra (typically ca 50 mg at 60 MHz). For coal-
derived samples containing chloroform-insoluble material, pyridine-d_5 can be
used but this solvent generally suffers from the difficulty in completely eliminating water and from low isotopic purity, which gives rise to interfering hydrogen
peaks. In order to avoid unwanted water and pyridine peaks, it is advisable to
use relatively large amounts of sample for FT measurements (ca >10 mg) with
standard 5 mm tubes (ca 0.5 cm^3 of solution) and to use a mixture of
chloroform-d with the smallest quantity of pyridine-d_5 to achieve solubility.

2.2.2 Spectra and peak assignments

Typical high-field spectra of a coal-derived oil and asphaltene fractions are
shown in Figure 2, which also indicates assignments for the principal bands. The
major separation is between aromatic and aliphatic hydrogen but distinct
regions can be identified within these bands, particularly for the oil. The much
superior resolution in the oil spectrum arises from the relatively low molecular
weight compounds present in the oil, e.g. phenanthrene ($H_{4,5}$ peak at ca 8.5
ppm), the protons of which probably have much longer T_2 values than those in
the high-MW species (ca 250–2000) comprising the asphaltenes.

The aromatic band in the asphaltene spectrum is broad and featureless but,
for the oil, peaks from both polynuclear aromatic compounds, such as phenanthrene, and single-ring aromatics can be detected, only the latter giving peaks
in the range 6.5–7.2 ppm. The chemical shifts of phenolic hydrogen (ArOH)
depend on the solvent and sample concentration, both of which influence the
extent of hydrogen bonding. Indeed, in Figure 2, the phenolic hydrogen peak in
the asphaltene spectrum overlaps the aromatic band and the difficulty in
identifying and measuring phenolic hydrogen by ^{1}H NMR has led to the use of
derivatization procedures (Section 2.3). In cases where a phenolic hydrogen
peak is visible between the aromatic and aliphatic bands, D_2O can be used to
exchange the phenolic hydrogen and remove the peak from the spectrum[17,18].
Olefins can be expected to occur in thermally cracked products and usually
give discernible peaks between 4.5 and 6.0 ppm. Bromine addition has been
used to confirm both the presence of olefins in coal pyrolysis tars (Figure 3)
and that CH_2=CHR is the predominant species present[17].

For aliphatic groups, the chemical shifts are governed principally by the
proximity of the hydrogen to an aromatic ring. In the earliest ^{1}H spectra of coal
extracts, it was observed that a separation between hydrogen adjacent (positioned α) to an aromatic ring and other aliphatic hydrogen[19] occurred at ca 2.0

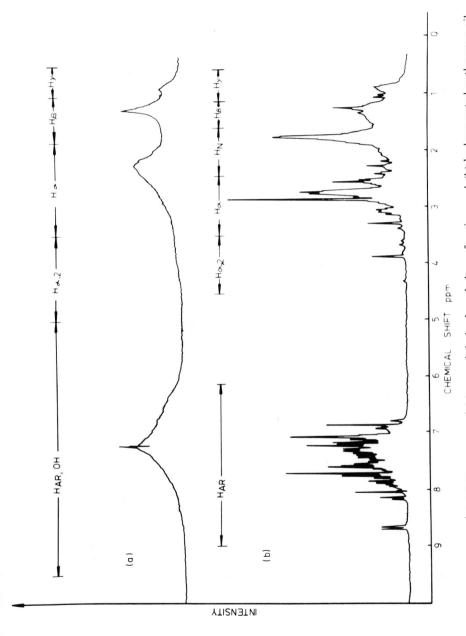

Fig. 2. 220 MHz ^{1}H NMR spectra of (a) a coal-derived asphaltene fraction and (b) hydrogenated anthracene oil

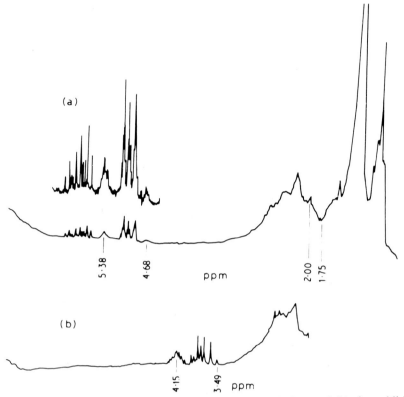

Fig. 3. 270 MHz ^{1}H NMR spectra of flash pyrolysis tar (a) before and (b) after addition of bromine[17]. Reproduced with permission of John Wiley and Sons

ppm. This separation can be observed in Figure 2 but occurs at slightly different chemical shifts for the oil and asphaltene fractions because of the vastly different proportions of aliphatic substituents present (see below). A shoulder at 3.4–4.5 ppm for the asphaltene fraction and distinct peaks in this range for the oil can be observed from hydrogen adjacent to two aromatic rings (e.g. CH_2 in fluorene, Figure 2, $H_{\alpha,2}$ region).

For petroleum and other products containing significant concentrations of alkanes, the most prominent peak in the 0–2 ppm spectral range occurs at 1.25 ppm, due to long-chain methylene protons. In the spectra of most samples at ca 1.0 ppm, there is usually a division between CH_2, CH not adjacent to an aromatic ring plus β-CH_3 (1–2 ppm) and any other CH_3 (0–1 ppm, paraffinic and that positioned γ or further from an aromatic ring). However, some CH_2 in cycloalkanes and naphthenic substituents may also contribute to the H_γ band. It can be seen in Figure 2 that only in the oil spectrum is there a prominent band between 1.5 and 2.0 ppm (H_N), this being attributable to β-CH_2 and CH in hydroaromatic compounds, such as tetralin. The presence of this band results in the separation between peaks from α- and other aliphatic hydrogen occurring at 2.1 ppm compared with ca 1.9 ppm for the asphaltene fraction where the concentration of α-hydrogen is much greater than that of the other aliphatic hydrogen (Figure 2).

The aliphatic hydrogen assignments discussed above are further complicated if there are significant concentrations of aliphatic groups adjacent to hetero-atomic groups, such as thioethers or thiols in petroleums and carboxyls and ethers in low-rank coal extracts. Peaks for $S—CH_n$ and $OOCCH_n$ ($n = 1$ or 2) occur between 1.8 and 3.0 ppm and consequently overlap those of protons adjacent to an aromatic ring (H_a), but OCH_n peaks occurs at significantly lower field (3.5–4.2 ppm)[20]. Clearly, some knowledge of elemental and functional group composition is desirable before attempting to assign the ^1H spectra of fuels.

The 2D COSY experiment has been particularly informative for oil samples. It has generally confirmed the broad assignment of aliphatic groups described above and has assigned unambiguously overlapping peaks due to C_1–C_4 alkyls, tetralins and indans in the spectra of monoaromatics from coal liquefaction oils[21] and diesel cuts[22] (Figure 4). CH_3 adjacent to an aromatic ring is not strongly coupled to other aliphatic groups and consequently no cross-peaks are observed in COSY spectra. Figure 4 indicates that peaks for α- and β-CH_2, CH in indans occur at slightly lower field than in their tetralin counterparts.

For oil samples, the dispersion of ^1H NMR spectra increases with increasing field strength. Figure 5 shows the aliphatic region for the 600 MHz spectrum of an aromatic fraction from a coal liquefaction distillate[23] and clearly the resolution is superior to that at 220 MHz (Figure 2), the addition of model compounds enabling individual CH_3 peaks to be assigned.

2.2.3 Quantification

Provided that a suitably long relaxation delay (>5 s) is used with a relatively small pulse angle ($<40°$) to acquire FT spectra and that correct phasing and flat baselines are achieved before integration (see Chapter 3), proportions of aromatic and aliphatic hydrogen can be determined with reasonable accuracy and precision. Indeed, recent work by the Institute of Petroleum Molecular Spectroscopy Panel in the UK[24] has demonstrated for a range of oils that the reproducibility of aromatic hydrogen contents (% aromatic H of total H or hydrogen aromaticity) is typically less than 1%, with the detection limit being less than 0.5% for lube oils. However, particularly for coal extracts, it should be ascertained whether or not phenolic hydrogen is included with the aromatic hydrogen (see above). Numerous workers have estimated the different hydrogen types that can be resolved by ^1H NMR ($H_α$, $H_β$, and $H_γ$), but the errors involved are probably greater than for aromatic hydrogen because only partial separation is achieved between the different bands even in high-field spectra.

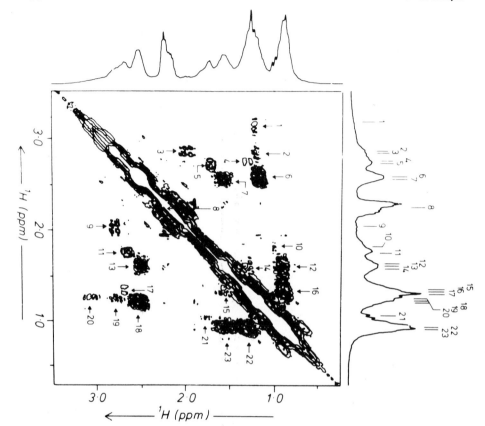

Fig. 4. Homonuclear correlation COSY 2D 200 MHz NMR spectrum of a monoaromatic shale oil diesel fraction shown as a contour plot. Cross-peaks have been interpreted as follows : 1–20, 1-methylindan ring; 2–19, isopropyl; 3–9, indian ring; 4–17, 1-methyltetralin ring; 6–18, ethyl; 7–(13/12)–23, propyl; 7–(13/14)–(15/16)–22, butyl; 8, α-CH$_3$ (not coupled); 10–21, 2-methyltetralin ring[22]. Reproduced with permission of the American Chemical Society

2.2.4 Some applications

A few of the numerous examples in which ^{1}H NMR has been used to give an indication of fuel structure without resorting to ^{13}C NMR are briefly summarized here. Oils used in coal liquefaction are most effective when relatively high concentrations of hydroaromatic compounds (hydrogen donors) are present. An indication of the concentration of hydrogen-donor groups can be obtained from the intensity of the H$_\alpha$ and H$_N$ bands[25,26] (Section 2.2.2). The aromatic hydrogen content of coal tars can be readily related to the pyrolysis conditions[11], increasing temperature giving more aromatic tars. Cetane

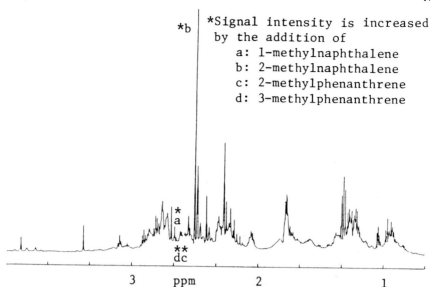

*b   *Signal intensity is increased
      by the addition of
          a: 1-methylnaphthalene
          b: 2-methylnaphthalene
          c: 2-methylphenanthrene
          d: 3-methylphenanthrene

Fig. 5. Aliphatic region of the 600 MHz ^{1}H NMR spectrum of the aromatic fraction from an SRC-II coal liquefraction oil[23]. Reproduced with permission of D. Riedel Publishing Company

numbers of diesel fractions have been predicted from the intensities of aromatic and aliphatic H_α, H_β and H_γ-bands[27].

^{1}H NMR is ideally suited for monitoring the compositions of chromatographic fractions from both aromatic ring size and molecular weight (GPC) separations (Figure 1). Effluents from HPLC can be analysed directly by ^{1}H NMR[28,29] but the necessity to use deuterated or non-protonated solvents may place limitations on LC separations. However, the direct coupling of LC and ^{1}H NMR is particularly useful for light cuts, such as diesel fuel[28,29], where problems with evaporation are encountered when recovering fractions from preparative separations.

2.3 ^{13}C Measurements

2.3.1   General

The first ^{13}C spectra of fuels were obtained by continuous sweep methods without broadband proton decoupling[9,30] However, since the introduction of commercially available FT spectrometers, virtually all spectra have been acquired with proton decoupling because of the obvious improvements in resolution although, initially, there were problems concerning quantification (Section 2.3.3).

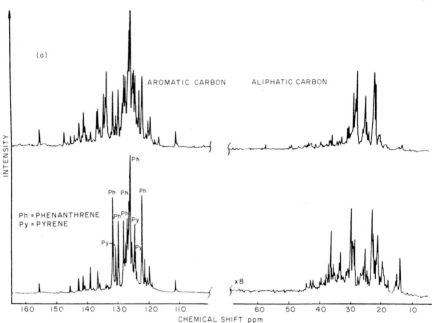

Fig. 6. 20 MHz ^{13}C NMR spectra of (a) initial and (b) recovered oil from a coal liquefaction experiment

^{13}C spectra of fuels are obtained routinely in chloroform-*d* because its chemical shift at 77.1 ppm falls conveniently between the aromatic and aliphatic bands (Figure 6). *sym*-Triazine has been proposed as a more powerful solvent than chloroform-*d* for primary coal liquefaction products[31] but its chemical shift at 170 ppm is close to the region for aromatic carbon bound to oxygen (148–165 ppm, see below). Although spectra of humic acids (sodium hydroxide-extractable material) from coals can be obtained in NaOD–D$_2$O, the resolution is vastly inferior to that achieved after methylation to render humic acids soluble in chloroform[32]. As an alternative to chemical modification to increase solubility (Section 2.1), ^{13}C spectra of coal liquefaction products containing chloroform-insoluble material have been obtained as melts at elevated temperatures[33], but the resolution does not match that achieved for solutions.

2.3.2 Spectra and peak assignments

Well resolved peaks are generally obtained in both the aromatic and aliphatic regions of proton-decoupled spectra of oils, even at low fields (Figure 6). Broader bands are usually found in the spectra of heavier fractions because the ^{13}C relaxation times are shorter (Section 2.3.3), but even asphaltene spectra

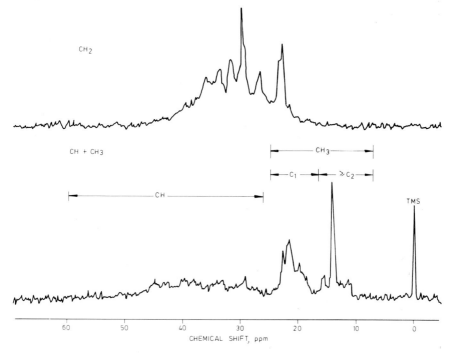

Fig. 7. 20 MHz aliphatic ^{13}C sub-spectra (obtained by the gated spin–echo method) for the asphaltene fraction from a coal hydrogenation product

contain sharp alkyl peaks (30 and 14 ppm in edited spectra, Figure 7). Assignments were based initially on reference spectra for suitable model compounds[34,35] before editing[36–39] and 2D methods[21,22,40,41] were implemented. Depending on the nature of the sample, a number of divisions in both the aromatic and aliphatic region can usually be made for fuel spectra.

The use of gated spin–echo[36–38] and DEPT methods[38,39] to resolve tertiary and quaternary aromatic carbon peaks has confirmed that a reasonable separation between mono- and diaromatic species occurs at ca 129.5 ppm (see Figure 8 for diesel fraction). However, the peaks for certain bridgehead and internal quaternary carbons in polyaromatic compounds, notably that for $C_{10b,c}$ in pyrenes at ca 124.5 ppm, occur at higher field than 129.5 ppm. Nonetheless, peaks for most bridgehead carbons in polyaromatic compounds occur in the range ca 129.5–132.5 ppm but those for diphenyls and fluorenes overlap the peaks for aliphatic-substituted quaternary carbon between 133 and 150 ppm[34]. The presence of heteroatoms, particularly oxygen groups, significantly extends the aromatic chemical shift range, peaks for C_{Ar}—O occurring between ca 115 and 120 ppm. In the spectra of asphaltenes and other heavy fractions, there is usually little resolution in the aromatic region, principally

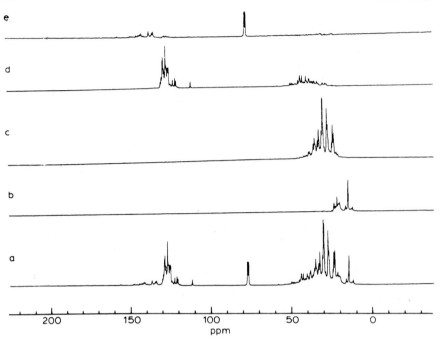

Fig. 8. ^{13}C NMR spectra of severely hydrotreated, coal-derived diesel oil[39]. (a) Normal spectrum obtained without NOE; (b) DEPT CH$_3$ sub-spectrum; (c) DEPT CH$_2$ sub-spectrum; (d) DEPT CH sub-spectrum; (e) quaternary only spectrum. Reproduced with permission of Butterworth Scientific Limited

because of the inherent complexity, although relaxation times are generally considerably shorter (and the signals consequently broader) than for oils (Section 3.2.3). If present in the spectra of shale oils and other pyrolysis products, olefin peaks can be resolved from overlapping aromatic peaks using the two-dimensional C/H correlation experiment since their ^{1}H chemical shifts are reasonably well separated[41] (Section 2.2.2). However, —CH=CH$_2$ end groups in alkyl chains give characteristic peaks at 139 (CH) and 114 ppm.

Spectral editing of aliphatic carbon bands by DEPT and gated spin–echo methods (Figures 7–9) has confirmed that there is a reasonable degree of separation between the chemical shift ranges for CH$_2$ (21–45 ppm) and CH$_3$ (10–24 ppm) even for asphaltene spectra. However, the ranges for CH (ca 25–50 ppm) and CH$_2$ overlap considerably and, before editing techniques were used, it was not possible to assign many CH peaks.

The 2D ^{1}H–^{13}C correlation experiment has confirmed that a number of divisions exist in the 10–24 ppm range for CH$_3$ peaks in the spectra of aromatic fractions[21,22] (Figure 10). There is overlap in the spectral region between 20 and

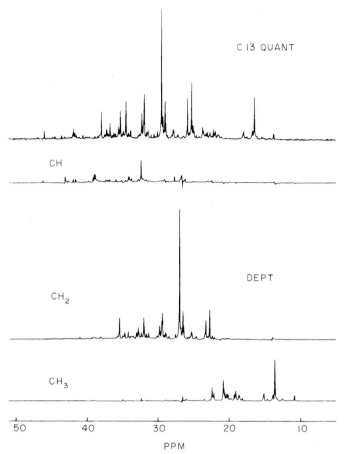

Fig. 9. 50 MHz aliphatic ^{13}C sub-spectra (obtained by the DEPT method) of a coal-derived oil[38]. Reproduced with permission of D. Riedel Publishing Company

24 ppm for peaks due to CH_3 substituted in aromatic and hydroaromatic/naphthenic rings (Figure 10) and, if present, isoalkyl groups. However, the chemical shift range for aryl-CH_3 groups not shielded by an adjacent group (20–22.5 ppm) occurs at higher field than for corresponding CH_3 groups adjacent to a second aromatic ring or substituent (e.g. in 1-methylnaphthalene)[34]. Ethyl-substituted aromatic groups give CH_3 peaks close to 16 ppm whereas peaks for CH_3 in larger alkyl substituents occur at higher field[21,22,34] (Figure 10), the prominent peak at 14 ppm in the edited and 2D spectra being attributable to the terminal CH_3 in $\geqslant C_3$ alkyl chains (Figures 7–10). The 10–15 ppm range in coal tars also includes peaks for CH_3 adjacent to two aromatic rings[34] (e.g. in 9-methylanthracene).

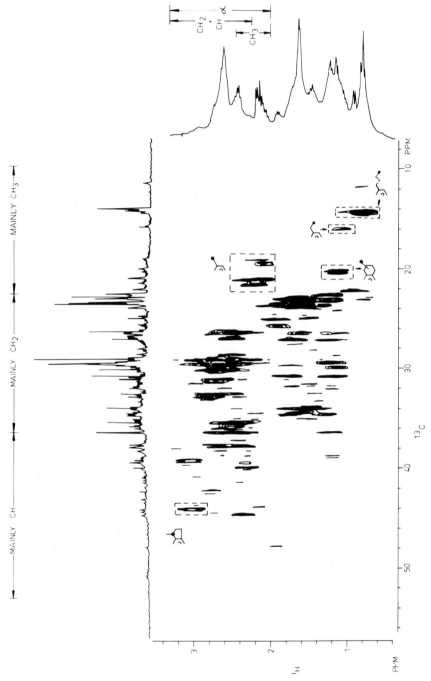

Fig. 10. $^{13}C/^1H$ chemical shift correlations for aliphatic groups in the monoaromatic fraction from a coal-derived oil (250–450 °C boiling range, ^1H frequency 300 MHz)

The most prominent peak in the aliphatic region of the spectra of petroleum and many coal liquid fractions occurs at 29.7 ppm[31,41] (Figure 8) and is attributable to CH_2 in long alkyl chains (positioned δ, ε or further from the end of the chain), related peaks occurring at ca 32 and 23 ppm (ν and β carbons). For high-boiling petroleum and some coal liquid fractions, these sharp alkyl peaks are superimposed on a broad unresolved shoulder in edited spectra (Figure 7) arising from CH_2 in hydroaromatic/naphthenic rings and alkyl CH_2 in close proximity (mainly in the α-position) to both aromatic and hydroaromatic/ naphthenic rings. These cyclic and alkyl CH_2 groups can be resolved by relaxation experiments, the more rapidly decaying component in T_2 experiments being attributable to cyclic carbon[42].

Isoalkyl CH, like long-chain alkyl CH_2, gives easily recognizable peaks at ca 28 and 33 ppm due to $-CH(CH_3)_2$ and $-CHR(CH_3)$ if present in significant amounts. However, the CH bands in edited spectra of oils are complex (Figures 8 and 9) because of the large chemical shift dispersion for alkyl-substituted and bridgehead CH in cyclic structures. Indeed, for heavy fractions, the CH sub-spectra usually are broad unresolved shoulders (Figure 7). Concentrations of quaternary aliphatic carbon in samples investigated by editing techniques have generally been insignificant[36-39]. The assignments discussed here have referred primarily to aliphatic substituents in aromatic molecules assuming that alkanes will be removed by LC prior to detailed NMR analysis. Although many alkanes can be readily identified by GC–MS, the identification of types of branched and cyclic groups is not straightforward and the use of edited ^{13}C spectra has enabled some of these groups to be identified and measured[43].

2.3.3   Quantitative aspects

It is now well established that quantitative ^{13}C spectra of oils and soluble fuel fractions can be obtained readily using gated decoupling to suppress nuclear Overhauser enhancements in conjunction with doping sample solutions with low concentrations of a non-interacting paramagnetic compound, such as chromium acetylacetonate[44-46] (see Chapter 3). Concentrations of 0.02–0.05 M of chromium acetylacetonate considerably shorten relaxation times and, conse-quently, also effectively suppress NOEs without causing appreciable peak broadening. However, for polar coal extract fractions and model phenols, it is generally observed that phenolic peaks ($C_{Ar}-OH$, 148–160 ppm) are broadened to the greatest extent, presumably due to complexation effects[46]. Also, if samples have to be recovered for other analyses, chromium acetylacetonate can only be separated from non-polar material (alkanes and aromatics) by LC. In practice, gated decoupling has to suffice for quantification. This increases accumulation times because relatively small pulse angles ($<40°$) with long relaxation delays (>10 s) must be used to ensure that complete thermal relaxation occurs. Quaternary aromatic carbons have considerably longer T_1 values than the aliphatic and tertiary aromatic carbons in fuels[47,48]. In fact, T_1 values decrease at higher field because of the contribution from chemical shift anisotropy to the relaxation rate[47]. Although T_1 values of fuels generally decrease with increasing sample viscosity, these are still typically 1 s for quaternary aromatic carbons in coal-derived asphaltenes[48].

The reproducibility of aromaticity determinations (proportion of aromatic carbon in total carbon) is typically 1% or less for oils[24] provided that appropriate experimental conditions are used and integrations have been carried out with flat baselines. The errors involved in deriving further information on aromatic and aliphatic groups by editing techniques and by arbitrary divisions of spectral regions are likely to be greater. A problem arises in DEPT and gated spin–echo methods for heavy fractions if T_2 values are short (< 40 ms) and there is a loss of signal intensity. Although cyclic carbons relax faster than alkyl carbons, a comparison of normal and gated spin–echo edited spectra for ashphaltenes suggested that the CH_3 concentration derived from the edited spectrum was not grossly overestimated[42]. There has been some discussion concerning the most appropriate editing technique for fuels[38,39,42]. Whereas DEPT easily resolves CH and CH_3 and does not require separate accumulations for aromatic and aliphatic carbon types[39], there is hardly any overlap between CH and CH_3 peaks for fuels[21,37]. Moreover, DEPT relies on polarization transfer and quaternary aromatic carbons cannot be measured directly, and in the DEPT experiment CH_2 intensity can be underestimated relative to CH and CH_3[39]. Clearly, inter-laboratory exercises would be useful for assessing the precision of editing methods for quantification. However, these methods are undoubtedly superior to that of indirect calculation procedures for estimating aliphatic H/C ratios and other important parameters (Section 4).

2.3.4   Some applications

Like ^1H spectra, ^{13}C spectra readily provide information on the changes in composition that occur for hydrogen-donor solvents during coal liquefaction (Figure 6) and can give an indication of the concentration of hydroaromatic groups[49,50]. The fact that long alkyl chains are easily identified by ^{13}C has enabled their concentration in coal tar and extract fractions to be estimated[51,52] and it was confirmed that long alkyl chain aromatics exist in coal liquids[51]. Structural calculations which combine data from ^1H and ^{13}C NMR are considered in Section 4.

2.4   Techniques for heteroatomic groups

Although non-aqueous titration methods exist for the estimation of concentrations of strongly acidic OH (phenols, carboxylic acids) and basic N groups (primary amines and aza compounds), there is a need to develop procedures for the measurement of neutral and weakly acidic O, N and S groups, such as ethers, furans, secondary amines, thioethers and thiophenes, particularly to gain a better understanding of the upgrading of heavy oils and bitumens.

2.4.1 Direct methods

^{17}O, ^{15}N and ^{33}S all suffer from low receptivity for fuels whereas ^{17}O, ^{14}N and ^{33}S are quadrupolar nuclei and often give broad signals. Not surprisingly, there have been few reports of the direct identification of heteroatomic groups in fuels by NMR although their chemical shift dispersions are large[53].

Figure 11 shows the ^{17}O spectra of two basic fractions separated from a coal liquefaction heavy distillate[54]. The signal-to-noise ratio is low despite ca 10^6 scans being accumulated by rapid pulsing. Nonetheless, peaks due to phenols (30–50 ppm), furans (ca 250 ppm) and possibly quinones (480–550 ppm) are discernible. The presence of the latter could be cross-checked by ^{13}C NMR (Section 2.3.2).

Sulphides are considered to be major form of sulphur in fuels but they give extremely broad ^{33}S peaks. However, when sulphides are oxidized to sulphones, much narrower peaks are obtained[55,56]. Moreover, there is a reasonably large chemical shift dispersion for sulphones. Unfortunately, both the author and other workers[57] have found that the ^{33}S spectra of heavy oil fractions contain only one unresolved band, presumably because of broadening originating from paramagnetic centres.

These preliminary results suggest that the direct observation of heteroatoms in fuels by multinuclear NMR will not play a major role in the foreseeable future, but the analysis of a range of heteroatomic-rich fractions would be desirable to assess fully the applicability of ^{17}O, ^{14}N, ^{15}N and ^{33}S NMR.

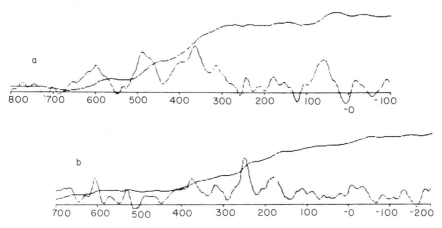

Fig. 11. ^{17}O NMR spectra of SRC-II heavy distillate fractions[54]. (a) Very weak bases; (b) strong bases. Reproduced with permission of D. Riedel Publishing Company

2.4.2   Indirect methods

Procedures in which labile hydrogens are replaced with groups containing a convenient magnetic label (^{1}H, ^{13}C, ^{19}F, ^{29}Si and ^{31}P) have been widely used to estimate OH groups (mainly phenolic and carboxylic), but NH, NH$_2$ and SH groups have also been identified and measured. As discussed above (Section 2.2.2), OH peaks are extremely difficult to measure by ^{1}H NMR because of overlap with aromatic and aliphatic hydrogen bands. Most of the procedures that have been applied to coal liquids and petroleums are summarized in Table 2[58-71]. The obvious advantage of a heteronuclear label is that there are no overlapping sample peaks but trimethylsilyl, acetyl and methyl derivatives all give well resolved peaks in either the ^{1}H or ^{13}C spectra (see below).

Reasonably reliable estimates of phenolic OH concentrations in primary coal liquefaction products have been obtained from the ^{1}H spectra of their trimethyl-silyl derivatives[58,59]. However, the ^{1}H and ^{13}C chemical shift dispersions for trimethylsilyl derivatives of phenols and alcohols (ca 1 ppm) are significantly smaller than that for ^{29}Si (Figure 12). In addition to ^{29}Si peaks of phenol and alcohol derivatives in the range 14–20 ppm[60,61], those of carboxylic acids occur at lower field (20–22 ppm) and when powerful reagents are used to derivatize weakly acidic aromatic secondary amines, such as indole and carbazole, peaks at much higher fields (<12 ppm) are observed. Although sharp peaks from individual phenols can be identified in the spectra of silylated high-temperature coal tar, silylated asphaltenes and other heavy fractions generally given broad unresolved bands. In Figure 13, the ^{29}Si spectrum of silylated asphaltenes from coal extract, a small band at 20–22 ppm due to carboxyl derivatives can be identified, the position and width of the major bank being consistent with phenols and naphthols being the major phenolic groups in the initial fraction. To ensure that ^{29}Si spectra are quantitative, it is advisable to use gated

TABLE 2
Summary of derivatization methods used in the NMR investigation of heteroatomic environments in fossil fuels

Method	Derivatives from phenols/alcohols	Other groups derivatized	Magnetic label	Ref.
Silylation	—OSi(CH$_3$)$_3$	COOH, NH	^{1}H, ^{29}Si	58–61
Hexafluoroacetone adduct formation	—OC(CF$_3$)$_2$OH	NH$_2$	^{19}F	62–64
Trifluoroacetylation	—OCOCF$_3$	NH, NH$_2$	^{19}F	65,66
Acetylation	—OCOCH$_3$	NH	^{13}C	58,67
Phosphorous-containing derivatives	—OPO(C$_2$H$_5$)$_2$, —OPS(CH$_3$)$_2$		^{31}P	68
Methylation	—OCH$_3$	COOH, NH$_2$, NH$_2$	^{13}C	69–71

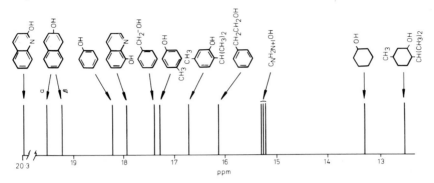

Fig. 12. ^{29}Si chemical shifts for trimethylsilyl derivatives of phenols and alcohols[60]

decoupling in conjunction with a paramagnetic relaxation agent as for ^{13}C NMR (Section 2.3.3).

Both trifluoroacetylation and the formation of hexafluoroacetone (HFA) adducts enable ^{19}F NMR to be used for the investigation of OH and other functional groups in fuels (Table 2). The ^{19}F chemical shift range of phenol –HFA adducts is ca 2 ppm and can be divided into regions due to adducts of

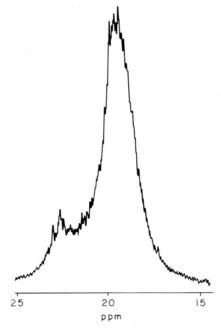

Fig. 13. 80 MHz ^{29}Si NMR spectrum of the trimethylsilylated asphaltenes from a coal extract

unhindered (*meta*- and *para*-substituted) and hindered (*ortho*-substituted) derivatives, the latter being deshielded by 0.4–1.2 pmm[62,63]. Moreover, the formation of HFA adducts can be carried out in an NMR tube and the formation constants for phenol–HFA adducts are large, with the exception of 2,6-disubstituted phenols[62]. In spectra of high-temperature coal tars, peaks due to individual phenol adducts (phenol, cresols and xylenols) can be identified and measured[63]. Not surprisingly, heavy coal liquids such as asphaltene fractions give a broad band, although shoulders due to adducts of hindered phenolic substituents can be identified[58]. However, estimates of phenolic OH concentrations for heavy coal liquids tend to be low[58], probably because hydrogen-bonding interactions in solution for polyfunctional molecules reduces the extent of HFA adduct formation. HFA forms adducts with primary amines and this has enabled anilines to be identified in coal-derived basic fractions[64]. Unfortunately, signals due to amine and phenol adducts overlap to a significant extent.

Trifluoroacetyl derivatives of phenols do not give as large a ^{19}F chemical shift dispersion as their HFA adduct counterparts[65]. Nonetheless, the use of trifluoroacetyl chloride enables a range of basic and weakly acidic compounds to be derivatized[65] (albeit not necessarily in high yield), and has enabled aminonaphthalenes to be identified in basic material from a coal liquefaction product[66]. The carboxyl peaks in the ^{13}C spectra of acetyl derivatives of phenols occur at ca 170 ppm and are reasonably well resolved from the aromatic carbon bands in both solution[58] and solid-state spectra[67] of coal extracts and coals, respectively. Acetylation in conjunction with ^{13}C NMR appear to give reasonably reliable estimates of phenolic OH concentrations for coal extracts[58] but, with acetic anhydride in pyridine, increasing the acetylation temperature causes aromatic secondary amines in addition to phenols to be derivatized.

Diethylphosphate and dimethylphosphinic esters of phenols have been prepared (Table 2) but the ^{31}P chemical shifts of *ortho*- and *para*-cresol derivatives are separated by only ca 0.5 ppm[68], which is probably not sufficient to provide information on the distribution of phenolic groups in heavier fractions.

Methylation in conjunction with ^{13}C NMR has probably been the most successful of the various approaches (Table 2) in differentiating OH and COOH groups[58,69] and in determining low concentrations of acidic groups in heavy petroleum fractions through the use of ^{13}C-enriched methyl iodide[69]. Carboxylic acids give methyl ester peaks close to 50 ppm whereas the peaks for methyl ethers of unhindered phenols occur at ca 55 ppm and those of alcohols and hindered phenols at ca 60 ppm[69,70]. The aliphatic regions from the ^{13}C NMR spectra of a petroleum residue before and after methylation with ^{13}C-enriched methyl iodide are shown in Figure 14[69]. The OCH$_3$ peaks for carboxylic acid and phenol derivatives (50–60 ppm) are well resolved from the aliphatic carbon band but there is severe overlap for NCH$_3$ and SCH$_3$ peaks. Nonetheless, the difference in the integral for the aliphatic carbon region after methylation allows estimates to be made of NH and SH concentrations in the parent residue[69]. In

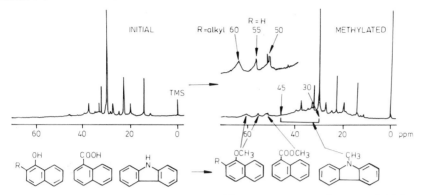

Fig. 14. Aliphatic carbon bands from the ^{13}C NMR spectra from an initial and methylated petroleum residue[69]. Reproduced with permission of the American Chemical Society

addition, ^{2}H NMR following methylation with CD_3 can be used to resolve SCD_3 (1–2.5 ppm) from NCD_3 and OCD_3 (2.3–5.0 ppm). Like acetylation, methylation had also been used in conjunction with solid-state ^{13}C NMR to investigate acidic functions in coals[71].

These examples demonstrate that derivatization–NMR procedures have not only been reasonably successful in estimating acidic OH concentrations but also in differentiating carboxylic acid and various phenolic OH environments which cannot be readily achieved by non-aqueous titration methods. Moreover, primary amines and weakly acidic NH and SH groups can also be identified.

2.5 ^{2}H NMR studies of coal liquefaction

The production of hydrogen is a major cost in coal liquefaction and clearly information on the fate of hydrogen introduced from either hydrogen-donor solvents or gaseous hydrogen is useful. Both deuterated tetralin and gaseous deuterium have been used in isotopic tracer experiments and the products analysed by deuterium NMR[72-77]. Figure 15 compares the ^{1}H and natural abundance ^{2}H spectra of a coal liquid, the bands being broader in the ^{2}H spectra of a coal liquid, despite the use of proton decoupling[76]. Gated spin–echo and INEPT methods have been used to identify the various tetralin isomers that arise after exchange between gaseous deuterium and unlabelled tetralin[77].

The results from studies with deuterium-labelled tetralin[72-75] have indicated that, at 400–450 °C, deuterium incorporation into coal-derived molecules is primarily in benzylic groups. However, at long reaction times, considerable scrambling of hydrogen and deuterium occurs with the incorporation of deuterium into aromatic structures and non-benzylic aliphatic groups.

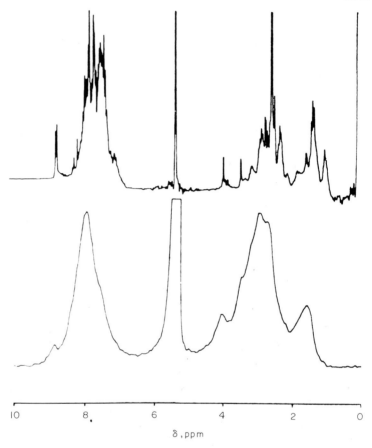

Fig. 15. 200 MHz ^{1}H (top) and 31.7 MHz ^{2}H (bottom) NMR spectra of the aromatic fraction of a lignite liquefaction product prepared with deuterium and carbon monoxide[76]. Reproduced with the permission of the American Chemical Society

3   SOLID-STATE AND LOW-RESOLUTION MEASUREMENTS

3.1 ^{13}C NMR

3.1.1   Techniques and spectra

Spectra of coals and oil shales, in which aromatic and aliphatic carbon bands are well resolved (Figure 16), can be obtained in reasonable accumulation times using the now well established techniques of (i) high-power decoupling to

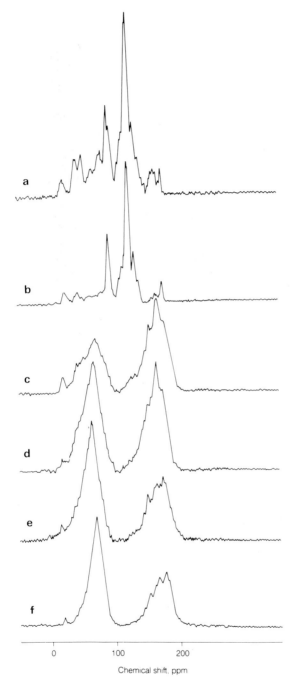

Fig. 16. ^{13}C CP–MAS NMR spectra (with ^{1}H dipolar decoupling) of peats (a,b), lignites (c,d,e) and bituminous coal (f)[78]. Reproduced with permission of Butterworth Scientific Limited

remove ^1H–^{13}C dipolar interactions generally in excess of 1 kHz; (ii) magic angle spinning (MAS) to remove chemical shift anisotropy (CSA), which is typically ca 100 ppm for aromatic carbons; and (iii) cross-polarization (CP) under Hartmann–Hahn conditions to increase the sensitivity and alleviate the need for long relaxation delays.

Although bands from carboxyl/carbonyl and OCH_n groups are resolved in the spectra of low-rank coals[78,79] (Figures 16 and 17), aromatic and aliphatic carbon bands in the spectra of solid fuels are usually broad and featureless. For coals, the broadening has been shown to arise mainly from the inherent complexity by using a selective saturation or 'hole burning' experiment[80]. Selective pulses to saturate ^{13}C spins over a relatively narrow frequency range (50 Hz) gave distinct minima or 'holes' at the saturating field frequencies in the resultant spectra; homogeneous broadening would have reduced the intensity of the aromatic and aliphatic envelopes without the appearance of minima. Indeed, inherent linewidths for a US coal have been estimated as ca 30–50 Hz from the ^{13}C T_2 values[80] (see below). Undoubtedly, in certain samples such as extraction[81] and hydrogenation[11] residues containing higher concentrations of organic free radicals and inorganic paramagnetic or ferromagnetic impurities, T_2 values are considerably shorter because demineralization greatly improves the resolution. The high moisture contents of lignites has been shown to degrade

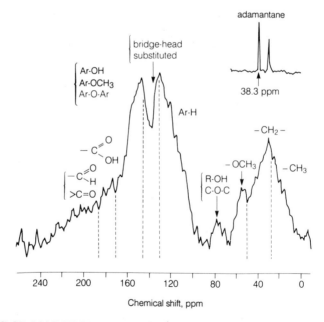

Fig. 17. ^{13}C CP–MAS NMR spectra (with ^1H dipolar decoupling) of Yallourn brown coal sample[79]. Reproduced with permission of Butterworths Scientific Limited

the resolution[82], since the high dielectric constant of water changes the sample capacitance and causes the probe to detune. Figure 18 demonstrates the improvement in both resolution and sensitivity achieved on drying for a typical lignite[82]. A thorough discussion of factors that can cause line broadening for coals can be found in Ref. 5.

The first published solid-state spectra of coal were obtained in 1976 without MAS[83] and showed significant overlap between the aliphatic and CSA-broadened aromatic bands. At low field ($\leqslant 25$ MHz), rotation speeds of ca 3 kHz are usually adequate to remove CSA and obtain sideband-free spectra. Obviously, much higher speeds are required at high field ($\geqslant 50$ MHz) and, although improvements in rotor design are making speeds well in excess of 10 kHz feasible, it has been necessary to resort to suppression techniques, such as TOSS (TOtal Suppression of Sidebands)[5,84,85] to obtain aromatic sideband-free spectra (Figure 19). However, the relative peak intensities may be affected by differences in ^{13}C spin–spin relaxation, since even at 3 kHz the 0.3 ms timescale of TOSS is considerable in relation to the T_2 values[5,80] (see above). At

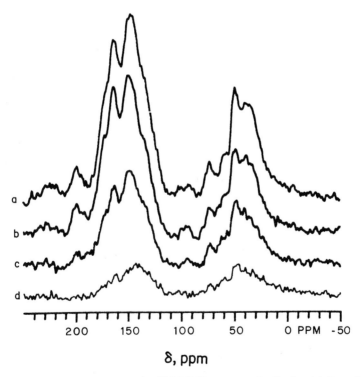

δ, ppm

Fig. 18. Effect of water content on the ^{13}C NMR spectra of a lignite. (a) Dry; (b) 6–7% water; (c) 13.0% water; (d) 22% water[82]. Reproduced with permission of the American Chemical Society

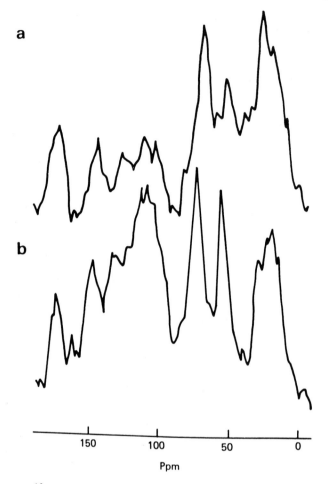

Fig. 19. 50 MHz ^{13}C NMR spectrum of a peat with MAS (rotation speed 3 kHz). (a) Normal spectrum; (b) TOSS spectrum. Reproduced with permission of Butterworth Scientific Limited

lower rotation speeds, aliphatic carbon signals are over-emphasized because these are more easily refocused than aromatic carbon signals with their broad distribution of sidebands[5,84]. It should also be mentioned that miscalculation of the delays in the TOSS sequence from variations in the rotation speed leaves significant sideband intensity in the resultant spectra[5].

The ^{13}C signal for many coals can be enhanced considerably by irradiating at the Larmor frequency of the electron. This technique is known as dynamic nuclear polarization (DNP) and its application to coals has been pioneered by workers at Delft University[86,87]. Two types of DNP experiments are employed:

either the ^{13}C spins are polarized directly by the free electrons or the ^1H spins are polarized first and normal CP is used to transfer the magnetization to the ^{13}C spins. Figure 20 compares spectra obtained by both these DNP methods and normal CP for a bituminous coal and shows the extremely high sensitivity attainable when DNP is used in conjunction with CP. It can be seen that the relative intensity of the aliphatic carbon band is lower in the spectra obtained with direct polarization of the ^{13}C spins (DNP–FID), presumably because the organic unpaired electrons (ca 10^{19} spins per gram in bituminous coals) are associated mainly with the aromatic groups. When the ^1H spins are polarized first (DNP–CP), rapid spin diffusion is thought to occur, giving rise to relative intensities of aromatic and aliphatic bands similar to those found in normal CP spectra. A large number of coals have been investigated by the DNP technique and compared with normal CP, ^{13}C signal enhancements of over 100 have been observed[86,87]; the enhancements are generally higher for high rank (bituminous) coals because of the greater organic free radical concentrations and the slower proton relaxation rates. However, a number of experimental parameters

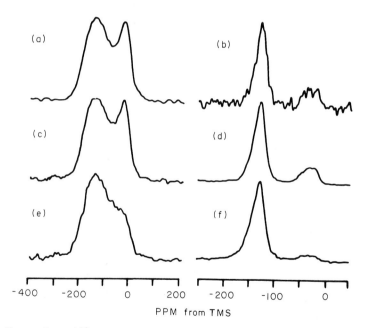

Fig. 20. Comparison of ^{13}C NMR spectra of a medium volatile bituminous coal obtained by different techniques[87]. (a) CP (contact time 0.9 ms, Lorentzian line broadening of 200 Hz, 20 000 scans); (b) CP–MAS (no broadening, 10 000 scans); (c) DNP–CP (400 scans); (d) DNP–CP–MAS (10 000 scans); (e) DNP–FID (90° pulse, 6 μs, 8 scans, recycle delay of 60 s, Lorentzian line broadening of 200 Hz); (e) DNP–FID–MAS (70 scans, no line broadening). Reproduced with permission of the American Chemical Society

such as the magnitude of the microwave field employed to irradiate the electrons, sample volume, degassing and powdering of coals affect the observed enhancement.

A number of approaches have been used to improve the resolution in the broad and featureless ^{13}C spectra of coals so that different aromatic and aliphatic groups can be identified and their concentrations estimated. Dipolar dephasing has been the most widely used and most successful technique for editing spectra according to carbon type[5,11,81,88–91]. If the high-power ^1H decoupling and ^{13}C irradiating fields are switched off for a period of ca 50 μs following cross-polarization but before data acquisition, the signal intensity of aliphatic CH_2 plus CH and aromatic CH is vastly reduced, leaving bands mainly due to quaternary aromatic carbon and highly mobile aliphatic groups, particularly CH_3 (Figure 21). This arises because the transverse magnetization decays much more slowly for the latter as a result of the much weaker ^1H–^{13}C dipolar interactions, typical decay constants being ca 15–30 μs for the CH_2 and CH groups and ca 100 μs for quaternary aromatic carbons (see Section 3.1.2). Two-dimensional dipolar dephased spectra have been obtained by varying the dephasing delay[92], quaternary and methyl carbons giving much narrower bands along the frequency axis for dipolar interactions.

Variable CP (contact) time and other relaxation experiments (Section 3.1.2) have also been used to resolve different carbon groups[5,11]. At short contact times (< 0.5 ms), quaternary carbons are discriminated against because they generally have the slowest CP rates, whereas at longer times (> 2 ms) for coals, peaks are not observed for ^{13}C spins adjacent to protons with short rotating frame relaxation times ($T_{1\rho}$, Section 3.1.2). One method that has not yet been explored is the differentiation of quaternary carbon, CH, CH_2 and CH_3 signals by removing ^1H–^1H dipolar interactions during spin–echo pulse sequences[93].

Aromatic carbon sidebands have generally been considered an unwanted intrusion in the high-field ^{13}C spectra of coals. However, bridgehead, alkyl-substituted and tertiary aromatic carbons have different chemical shielding tensors which give rise to different sideband intensities. Therefore, their intensities can be used to give an indication of the proportions of these aromatic carbon types and it has been shown with this approach that anthracite contains extremely high concentrations of bridgehead carbons[94]. The use of sideband analysis in conjunction with Lorenzian to Gaussian enhancement[95] has been particularly successful for assigning aromatic carbon peaks in low-rank coal spectra. For example, from the sideband intensities a peak at 144 ppm was assigned to carbon adjacent to two oxygen groups.

Perhaps the boldest approach adopted to extract quantitative information from the broad aromatic and aliphatic carbon bands in solid-fuel spectra has been the use of peak synthesis[96]. Chemical shifts for model compounds have been used to deduce peak widths and shapes for a number of aromatic and aliphatic carbon types. These peaks are then taken to synthesize the aromatic

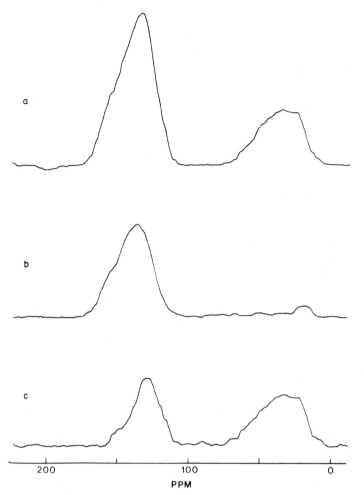

Fig. 21. ^{13}C NMR spectra of an Illinois No. 6 vitrain. (a) Normal spectrum (1 ms contact time); (b) quaternary carbon and mobile aliphatic carbon sub-spectrum; (c) difference spectrum (a − b)[90]. Reproduced with permission of Verlag Glückauf GmbH

and aliphatic bands found in actual spectra. The carbon distribution for an oil shale has been reported[96], but it will be interesting to compare solid-state and spectral editing solution-state results for soluble fractions, such as asphaltenes.

3.1.2 Quantification

There has been considerable debate over whether ^{13}C spectra of coals can be considered quantitative. As for other materials (see Chapter 3), the optimum conditions for obtaining quantitative CP spectra are

$$^1H \ T_{1\rho} \gg \text{contact time} \gg T_{CH}$$

where T_{CH} = CP time constant. Although values of T_{CH} for quaternary carbons are considerably larger than for most CH_n groups[5,97], the principal problems for coals arise from the presence of organic free radical centres. These are responsible for the relatively short $^1H \ T_{1\rho}$ values; coals generally display multi-component behaviour with typically up to 50% of the ^1H spins in bituminous coals having $T_{1\rho}$ values less than 1 ms[98,99] (the exact values are dependent on the strengths of both the static and spin-locking fields), which is not long enough for complete cross-polarization. Unfortunately, most $^1H \ T_{1\rho}$ measurements have been made indirectly by modifying the basic CP sequence[5,80] and this fails to detect the faster relaxing components. Carbons in the vicinity of free radicals are not likely to be observed because of their extremely short T_2 values. Clearly, this discussion suggests that much of the carbon in coals is not observable by CP ^{13}C NMR. Indeed, estimates by internal standard[100] and carbon counting[101] procedures suggest that only about half is measured. The single pulse or 90°-Block decay experiment generally allows more of the carbon to be sampled[100] and, for some coals, higher aromaticity values have been deduced[102] by this method than by CP. However, the signal to noise levels in Block decay experiments are extremely poor because of the long ^{13}C thermal relaxation times. Recycle times of at least 10 s are required to maximize signal intensity and not underestimate aromaticity values due to long T_1 value of quaternary aromatic carbons (Figure 22).

Soluble extract fractions, such as asphaltenes, are generally inappropriate for investigating quantitative aspects for coals[99]. The principal difference is that ^1H $T_{1\rho}$ values are considerably larger (> 10 ms) in solution, with the consequence that much more of the carbon is observed and aromaticity estimates from solid and solution state spectra are in close agreement[11,99]. However, for soluble materials, rapid molecular motion may reduce ^1H–^{13}C dipolar interactions, making CP slower. Also, $^1H \ T_1$ values are generally considerably larger than those for coals[5,99] (ca 100 ms) and longer recycle times are needed. This problem can be alleviated with the use of a flip-back pulse after CP to help re-establish the ^1H magnetization[5,99].

Neglecting the factors outlined above that can affect relative peak intensities, the estimated error in measuring the aromaticity of a bituminous coal from a solid-state ^{13}C spectrum is generally considered to be $\pm 2\%$ of the total carbon[89,103]. As an alternative to selecting an optimum contact time for CP spectra (usually ca 1 ms to achieve maximum signal intensity and a relatively high degree of CP for quaternary carbons), aromaticity values have been deduced from variable contact time experiments by fitting the aromatic and aliphatic carbon magnetization curves to the approximate form

$$M_t = M_0(1 - \exp[t/T_{CH}]) \exp[-t/(^1H \ T_{1\rho})]$$

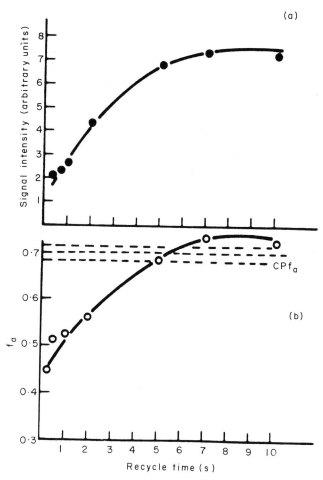

Fig. 22. Effect of varying the recycle time on (a) total signal intensity and (b) measured aromaticity (f_a) in the Block decay experiment on an Australian coal (the f_a value obtained from a CP experiment is also included[11]). Reproduced with permission of Pergamon Press

In other words, the magnetization increases at a rate of T_{CH}^{-1} and decreases at a rate of $(^1H\ T_{1\rho})^{-1}$. This approach compensates for the slower CP rate of quaternary aromatic carbons.

Plots of aromaticity values derived from CP spectra vs H/C ratio for a range of coals[11,99] show considerable scatter which the author considers to arise as a consequence of not all the carbon being sampled. Indeed, it would appear that representative carbon samples are obtained for some coals, since good agreement is achieved for aromaticities derived from CP and Block decay experiments[11,80] but not for others.

Concentrations of quaternary aromatic and mobile aliphatic carbon can be estimated by fitting the magnetization curves from dipolar dephasing experiments carried out with variable dephasing delays (Section 3.1.1) to an equation of the form

$$I_t = I_A \exp\left(-0.5t/T_{2A}\right)^2 + I_B \exp\left(-t/T_{2B}\right)$$

where t = dipolar dephasing time, I_A and I_B = initial intensity for faster and slower decaying components, respectively, and T_{2A} and T_{2B} = time constants for the two decays. The slower decaying component is Lorenzian and the faster one is Gaussian (second-order exponential). However, Figure 23 demonstrates that unusual decay curves can arise through the magnetization intensity being subjected to rotational and dipolar modulations[88]. Clearly, the results from dipolar dephasing experiments also depend on the factors discussed above that can affect peak intensities in CP spectra.

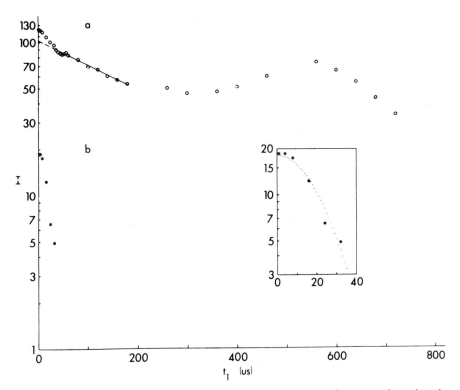

Fig. 23. (a) Plot of ln(peak intensity) vs dipolar dephasing time for aromatic carbon in bituminous coal (the solid line is the least-squares line of best fit); (b) calculated data for the rapidly decaying component (the dotted line represents the least-squares curve of best fit)[88]. Reproduced with permission of the American Chemical Society

3.1.3 Some applications

Despite the uncertainties in quantification, there are numerous important applications of CP ^{13}C NMR in fuel science. Oil yields from the pyrolysis of oil shales have been found to correlate closely with the amount of aliphatic carbon present[104,105] (Figure 24).

A wide range of peats, brown coals, lignites, bituminous coals and anthracites have been examined and the loss of OCH_n (cellulose), carboxyl and aliphatic carbon observed as coalification progresses[10,78,106]. Coal is extremely heterogeneous, consisting of visibly distinct microscopic domains known as macerals and CP spectra have confirmed the wide variation in aromaticity that exists between the different macerals[10,89,92,107]. Both dipolar dephasing[91] and spinning sideband analysis[94] have indicated that anthracites consist of highly condensed aromatic structures compared with bituminous coals. The use of

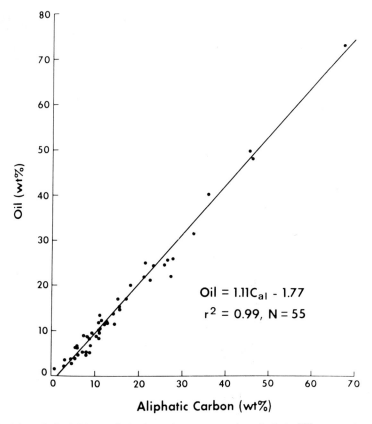

Fig. 24. Plot of oil yields vs aliphatic carbon content for oil shales[105]. Reproduced with permission of D. Riedel Publishing Company

solution- and solid-state ^{13}C NMR has enabled the net changes in aromaticity during coal liquefaction to be evaluated[108,109] and the results suggest that some hydrogenation of coal occurs at 400 °C in hydrogen at high pressures. Analysis of ^{13}C-labelled products from alkylation[110] and artificial coalification[111] reactions has helped to elucidate coalification pathways. A detailed account of these and other applications of CP ^{13}C NMR in coal science has recently been published[10].

3.2   High-resolution ^{1}H NMR

The combination of MAS and high-power multi-pulse methods, such as MREV and BR24[5,112], to remove homonuclear aliphatic interactions has enabled solid-state ^{1}H NMR spectra in which the aromatic and aliphatic bands are partially resolved[98,113] to be obtained. Combined rotation and multi-pulse spectroscopy (CRAMPS) is particularly informative for brown coals (Figure 25) with methoxy, carboxyl and phenolic peaks being resolved. The CRAMPS spectra of two coal-derived asphaltene fractions are shown in Figure 26 and it can be seen that the resolution is superior to that obtained for bituminous coals[98], presumably because the concentration of free radicals is much lower[99]. Encouragingly, the intensities of the aromatic hydrogen bands appear to be broadly similar to those in the solution-state spectra (Figure 26).

The use of CRAMPS in conjunction with an appropriate high-resolution ^{13}C pulse sequence[114] has enabled an elegant 2D ^{1}H–^{13}C chemical shift correlation to be carried out on a bituminous coal (Figure 27). In the projections perpendicular to the ^{1}H axis, aliphatic bands due to CH_2 and CH_3 can be resolved. This method clearly has considerable potential for probing differences between coal macerals and structural changes that occur during coalification (Section 3.1.3).

3.3   Low-resolution techniques

The determination of the hydrogen contents of liquid fuels at low field with the Newport analyser (Oxford Instruments)[12,115] undoubtedly represents the major routine use of NMR in fuel analysis. Indeed, this simple continuous wave analyser can be used to determine the moisture contents of coals and the composition of coal/oil slurries[116], since relatively sharp signals are observed for free water and for oil.

Pulsed ^{1}H methods have been used to probe coal structure and to investigate the pyrolytic behaviour of coals, pitches and oil shales. The transverse magnetization free induction decays of coals have been resolved into fast and slowly decaying components[117], T_2 values being 12 and 35 μs for a bituminous coal. It

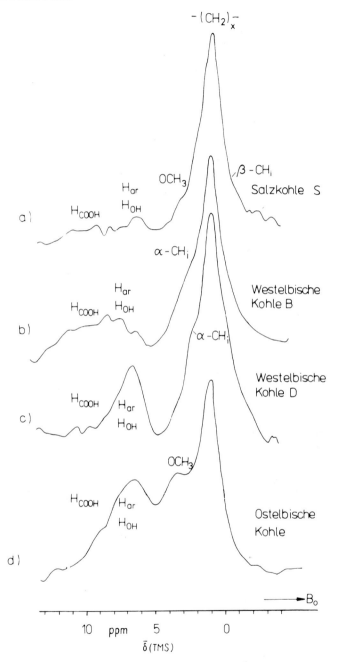

Fig. 25. High-resolution 270 MHz ^{1}H NMR spectra of German brown coals[113]. Reproduced with permission of the authors

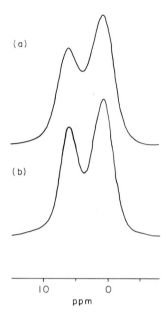

Fig. 26. High-resolution 300 MHz solid-state ^{1}H NMR spectra of two coal-derived asphaltene fractions. (a) Aromatic plus phenolic hydrogen content ca 30% (b) aromatic plus phenolic hydrogen content ca 42%

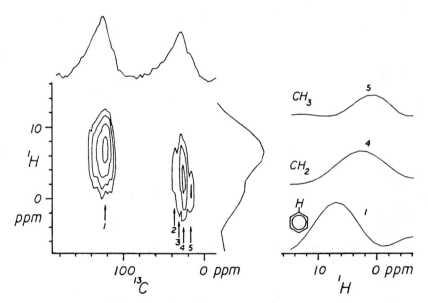

Fig. 27. Contour plot of the H/C heteronuclear shift correlation for an Illinois No. 6 coal[114]. Reproduced with permission of the American Chemical Society

is generally agreed that the two components arise from differences in mobility[118], but it is probably too simplistic to assume that the mobile (faster relaxing) component arises solely from molecules held within the macromolecular framework of coals. However, extraction of coal with pyridine removes most of the mobile hydrogen[119]. On the other hand, swelling coals in deuterated pyridine significantly increases the proportion of mobile hydrogen[119,120] and gives rise to more complex FIDs, possibly due to increased mobility in the macromolecular structure.

High-temperature probes operating at up to 600 °C have enabled both changes in total hydrogen content (corrected for increasing temperature) and mobility to be detected during pyrolysis of coals, pitches and oil shales[13,121,122]. The results from a typical thermal analysis experiment on a maceral concentrate from coal are shown in Figure 28. The power spectrum second moment (M_2^*) is a qualitative indicator of the average mobility[13,121] and is derived from FIDs obtained with the solid-echo pulse sequence ($90°_x$-T-$90°_y$) which is used to alleviate problems with receiver dead time. In Figure 28, the thermoplastic region is indicated by the minimum in M_2^* (ca 430 °C). Interestingly, devolatilization, as indicated by the decreasing hydrogen content, continues beyond the loss of plasticity. Cooling the pyrolysed coal gives little change in M_2^*, confirming the irreversibility of the changes that have occurred. In contrast to coals, the linewidths for fluid materials, such as coal tar pitch and coal extract, remain relatively small over a wide temperature range[122].

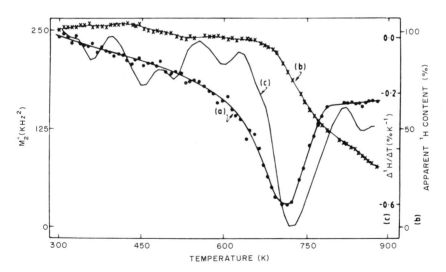

Fig. 28. Pyrolysis of vitrinite concentrate of an Australian high-volatile bituminous coal at 4 K min^{-1}. (a) Power spectrum second moment M_2^*; (b) apparent hydrogen content; (c) differential hydrogen content (arrows indicate values measured during cooling)[13]. Reproduced with permission of D. Riedel Publishing Company

4 STRUCTURAL ANALYSIS

4.1 General

^{1}H and ^{13}C NMR results for heavy oil and coal liquid fractions are often combined with elemental compositions, number-average molecular weights and concentrations of heteroatomic groups, usually with some assumptions, to derive more detailed information on the nature of aromatic and aliphatic substituents. Structural analysis results have been presented in three forms: (i) structural parameters[123-126], such as the average size of aliphatic groups and number of aromatic rings (see Table 3 for a typical set; the symbols are different to those used in the text, reflecting the lack of standardization); (ii) hypothetical average structures that closely fit the calculated parameters; and (iii) concentrations of particular groups[6,127] (Table 4), a relatively small number being used to describe the complex molecular compositions of heavy oils.

TABLE 3
Set of typical structural parameters used to characterize heavy coal liquid and petroleum fractions[123]

Symbol	Average parameter	Equation
n	Carbons per alkyl side-chain	$\dfrac{H_\alpha + H_\beta + H_\gamma}{H_\alpha}$
f_c	Carbon: hydrogen weight ratio of total alkyl groups	$\dfrac{CC_{\text{ali}}}{HH_{\text{ali}}}$
x	Hydrogen: carbon atomic ratio of alkyl groups	$\dfrac{12}{f_c}$
No. C_A	Aromatic carbons per average molecule	$\dfrac{C_A M}{1200}$
No. C_1	Aromatic non-bridge carbons per average molecule	$\dfrac{C_1 M}{1200}$
R_A	Aromatic rings per average molecule	$\dfrac{\text{No. } C_A - \text{No. } C_1}{2} + 2$
A_S	Percentage substitution of aromatic rings	$\dfrac{100 C_1^S}{C_1}$
R_S	Alkyl substituents per average molecule	$\dfrac{AS \text{ No. } C_1}{100}$
R_N	Naphthenic rings per average molecule	$R_S(n + 0.5) - \dfrac{6n}{f_c}$
No. C_{ali}	Aliphatic carbons per average molecule	nR_S
No. H_{ali}	Aliphatic hydrogens per average molecule	$n \times R_S$

The last approach was devised to alleviate the main problem of structural parameters and average structures, which is that little information on the distribution of structures present is derived. The accuracy of structural calculations is obviously dependent on the accuracy of the NMR and other data used but also on the assumptions made and both random and systematic errors can easily arise[128]. Nonetheless, structural analysis procedures have generally improved our understanding of coal liquefaction and heavy oil upgrading (Section 4.4). The examples discussed here are limited to soluble products, although solid-state ^{13}C NMR data can also be used in structural analysis[89], but the errors are probably greater because of uncertainties concerning quantification (Section 3.1.2). Much of the early work on structural analysis was

TABLE 4
Groups used to define the compositions of heavy coal liquid and petroleum fractions[6,127]

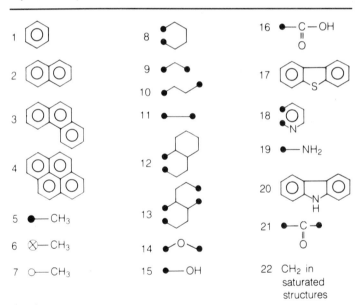

notation:

⬤— bound directly to an aromatic ring

◯— bound to a carbon alpha to an aromatic ring

⊗— bound to a carbon beta or further from an aromatic ring

example:

concerned with estimating aromaticities indirectly from ^{1}H NMR data and C/H ratios by the equation initially used by Brown and Ladner[19]:

$$f_a = \frac{C/H - H_\alpha/x - H_\beta/y}{C/H}$$

where H_α and H_β are proportions of aliphatic hydrogen adjacent and not adjacent to an aromatic ring, respectively, and x and y are the assumed aliphatic H/C ratios for these environments. Systematic errors can arise from the choice of aliphatic H/C ratios and, because aromaticities are now routinely determined by ^{13}C NMR, this indirect calculation procedure is not considered further.

4.2   Aromatic groups

The objective in most schemes[124–126] has been to deduce the proportions of bridgehead or internal aromatic carbon (i.e. that bonded only to other aromatic carbons) from which the degree of condensation of aromatic structure can be gauged. Although signals from individual *cata-* and *peri-*condensed aromatics can be identified in the ^{13}C spectra of coal oils (Section 2.3.2), there is usually little resolution for heavier fractions. Thus, concentrations or numbers of bridgehead aromatic carbons (C_{INT}) have been deduced from summing all the peripheral aromatic carbons (C_p), i.e.

$$C_{INT} = C_{AR} - C_P$$

$$C_P = H_{AR,OH} + C_\alpha + 2RJG$$

where $H_{AR,OH}$ is the number of aromatic and phenolic hydrogens, C_α is the number of carbons adjacent to the aromatic ring and RJG is the number of ring-joining groups, ArXAr structures (where X $= CH_2/CH$, O, NH or S).

Whereas the $H_{AR,OH}$ term can be deduced from ^{1}H NMR, ^{13}C spectral editing and measurements of OH concentrations, significant errors are likely to arise for the C_α and RJG terms. C_α is deduced from the proportion of aliphatic hydrogen adjacent to an aromatic ring by assuming that the aliphatic H/C ratio for α-groups is the same as that for all the aliphatic groups. However, there is usually only a partial separation between α-hydrogen and other aliphatic bands in ^{1}H spectra of fuels (Section 2.2.2).

A knowledge of heteoatom environments is essential for calculating the RJG term. However, for petroleum fractions which usually contain relatively high sulphur concentrations, the RJG term has not been included in the periheral aromatic carbon calculations[124,128], with the result that C_{INT} values are overestimated. Moreover, sulphur may occur in aliphatic thioethers and thiophenes, which would also affect the C_α terms. Clearly, it is beneficial to fractionate heavy products in order to limit the number of structural possibilities as far as heteroatomic environments are concerned (Figure 1). Aromatic material separated from n-pentane/n-heptane-soluble material (neutral oils) generally contains low concentrations of heteroatoms and, by definition (non-acidic and non-basic functions), these should be in ring-joining positions. For asphaltenes and other heavier fractions, heteroatom concentrations are usually high, particularly for coal extracts and primary liquefaction products[129]. However, oxygen is usually the major heteroatom present and, for bituminous coal products, non-phenolic oxygen can safely be assumed to be mainly in aromatic ether/dibenzofuran structures.

Because of the uncertainties in calculating C_{INT}, confirmatory evidence from other techniques is desirable. The author has found that values for a series of coal tar and extract fractions are consistent with results from differential pulse voltammetry[130], the characteristic reduction potentials for aromatic nuclei giving an indication of the distribution of structures present. For the neutral oil fractions, LC separations should also give information on the range of aromatic structures present (Figure 1).

4.3   Aliphatic groups

The parameters discussed here refer to aliphatic substituents bonded to aromatic groups but, as mentioned previously, paraffinic fractions can be analysed in detail by ^{13}C NMR[43]. Indeed, the use of ^{13}C spectral editing methods[36,37,42], in particular, has enabled much more information to be derived directly from NMR (Section 2.3).

The average number of carbons per substituent (n in Table 3) can be deduced by dividing the total aliphatic hydrogen content by the proportion adjacent to an aromatic ring (Table 3), hydroaromatic groups counting as two substituents. It has long been recognised that petroleum fractions generally contain much larger substituents than corresponding coal-derived fraction[131].

Aliphatic H/C ratios and the number of hydroaromatic/naphthenic rings per molecule (R_N) have traditionally been derived indirectly from C and H contents and proportions of aromatic and aliphatic groups[123,124,128]. (Table 3). However, these calculations procedures are prone to error[37,128]. Highly condensed naphthenic groups have been proposed as the major aliphatic substituents in primary coal liquefaction products[132] on the basis of grossly underestimated aliphatic H/C ratios (1.3–1.7). The gated spin-echo ^{13}C editing method has in fact shown that aliphatic H/C ratios for coal liquids occur in the approximate range 1.9–2.4, with short alkyl substituents accounting for much of the aliphatic carbon in pyrolysis products[51].

The critical assessment of structural analysis by Shenkin[128] highlighted the lack of precision in calculating the number of naphthenic rings per molecule. The estimation of the proportions of alkyl and naphthenic carbons can be best tackled by the identification of characteristic alkyl and isoalkyl peaks in CH, CH_2 and CH_3 sub-spectra[21,42,43]. For heavy fractions, such as asphaltenes, ^{13}C T_1 and T_2 experiments should enable alkyl CH_2 in close proximity to aromatic and naphthenic rings to be resolved from cyclic carbon[42]. Also, the use of dehydrogenation in conjunction with ^{13}C NMR should prove useful for characterizing naphthenic substituents in heavy fractions not amenable to GC–MS.

4.4 Some applications

It is apparent from the foregoing discussion that some of the structural variations reported in the literature could well reflect inherent deficiencies in the calculation procedures used. Nonetheless, particularly for heavy oils and primary coal liquefaction and pyrolysis products, structural analysis has proved to be informative.

High-yield extracts from coals are considered to be reasonably representative of the parent coal structure and those from bituminous coals containing ca 80 % carbon (dry ash-free basis) have been found to consist primarily of one and two ring aromatic structures[126,130]. With increasing processing severity, tars and extracts become more aromatic in character and contain more highly condensed aromatic nuclei[17,126,133,134]. It has been possible to define petroleum and coal-derived asphaltenes in terms of average structural properties using the proportion of bridgehead aromatic carbons (C_{INT}/C) to give an indication of π–π associative interactions in conjunction with number-average molecular weights and acidic OH concentration[129], the latter providing a measure of hydrogen-bonding interactions.

In order to gain some insight into the distribution of structural parameters about their statistical average values in heavy fractions, NMR has been used to characterize sub-fractions separated by size-exclusion chromatography (Figure 1). Wide variations in aromaticity and the average size of aliphatic substituents have been found with increasing molecular weight[135–137].

The use of the group composition approach[6,127] (Table 4) has enabled thermodynamic properties, such as heat capacity, to be predicted for coal liquids[138,139]; the predicted variation in heat capacity with increasing temperature has been found to be in close agreement with experimental results.

5 CONCLUSION

It is hoped that the pre-eminent role played by NMR in the structural characterization of fossil fuels has been conveyed to the reader. For established solution-state ^{1}H and ^{13}C measurements correlations with structure, formation and conversion are continually being found. Broadline ^{1}H methods are also proving useful in probing coal and oil shale pyrolysis and the macromolecular structure of coal.

It is difficult to predict the impact that newly emerging NMR techniques will have on the characterization of fuels. However, the resolution achievable in high-field imaging (ca 10 μm spatial resolution) should make it feasible to probe coal structure. Incorporation of deuterium into solid fuels should permit zero-field spectra to be obtained in which the resolution is comparable to that in high-field ^{2}H solution spectra[114]. Double cross-polarization techniques[140] (e.g. ^{1}H–^{31}P–^{13}C) may facilitate the ^{13}C analysis of solid fuels in which the spectra contain peaks for carbons adjacent to specific sites. The only certain prediction is that fossil fuels will continue to present a formidable challenge for new NMR techniques.

REFERENCES

1. B. P. Tissot and D. H. Welte, *Petroleum Formation and Occurrence*, Springer, Berlin, Heidelburg, New York, 1978.
2. P. H. Given. In M. L. Gorbaty, J. W. Larsen and I. Wender, (eds), *Coal Science*, Vol. 3, Academic Press, New York, 1984, p. 63.
3. K. H. Altgelt and T. H. Gouw (eds), *Chromatography in Petroleum Analysis*, Marcel Dekker, New York, 1979.
4. K. Van Nes and H. A. Van Westen, *Aspects of the Constitution of Mineral Oils*, Elsevier, New York, 1951.
5. D. E. Axelson, *Solid State Nuclear Magnetic Resonance of Fossil Fuels*, Multiscience, Edmonton, Canada, 1985.
6. L. Petrakis and D. Allen, *NMR for Liquid Fossil Fuels*, Elsevier, Amsterdam, New York, 1986.
7. R. B. Williams, *ASTM Spec. Tech. Publ.*, No. 224, 168 (1957); *Spectrochim Acta*, **14**, 24 (1959).
8. H. S. Rao, G. S. Murti and A. Lahiri, *Fuels*, **39**, 263 (1960).
9. R. A. Friedel and H. L. Retcofsky, *Chem. Ind. (London)*, 455 (1966).
10. R. M. Davidson, *Nuclear Magnetic Resonance of Coal*, ICTIS/TR32, IEA Coal Research, London, 1986.
11. M. A. Wilson and A. M. Vassallo, *Org. Geochem*, **8**, 299 (1985).
12. *ASTM Method D3701-78*, 1988 Annual Book of ASTM Standards, American Society for Testing and Materials, Philadelphia, 1978.
13. L. J. Lynch, D. S. Webster, N. A. Bacon and W. A. Barton. In L. Petrakis and J. P. Fraissard, (eds), *Magnetic Resonance—Introduction, Advanced Topics and Applications to Fossil Energy*, Reidel, Dordrecht, 1984, p. 617.

14. C. E. Snape and K. D. Bartle, *Fuel*, **58**, 898 (1979).
15. R. A. Grienke, *Fuel*, **63**, 1374 (1984).
16. A. N. Twigg, R. Taylor, M. K. Marsh and G. Marr, *Fuel*, **66**, 28 (1987).
17. P. J. Collin, R. J. Tyler and M. A. Wilson. In H. D. Shultz (ed.), *Coal Liquefaction Products, Vol. 1, NMR Spectroscopic Characterisation and Production Processes*, Wiley, New York, 1983.
18. K. D. Bartle, W. R. Ladner, T. G. Martin, C. E. Snape and D. F. Williams, *Fuel*, **58**, 413 (1979).
19. J. K. Brown and W. R. Ladner, *Fuel*, **39**, 87 (1960).
20. C. J. Pouchart and J. R. Campbell, *Aldrich Library of NMR Spectra*, Aldrich, Milwaukee, WI, 1974.
21. C. E. Snape, G. J. Ray and C. T. Price, *Fuel*, **65**, 877 (1986).
22. D. J. Cookson and B. E. Smith, *Energy Fuels*, **1**, 111 (1987).
23. T. Hara and N. C. Li. In L. Petrakis and J. P Fraissard (eds), *Magnetic Resonance—Introduction, Advanced Topics and Applications to Fossil Energy*, Reidel, Dordrecht, 1984, p. 729.
24. N. Evans, *Fuel*, **66**, 438 (1987); Proposed Method, 1989 IP Handbook (in press).
25. F. P. Burke, R. A. Winschel and T. C. Pochapsky, *Fuel*, **60**, 562 (1981).
26. D. J. Cookson, J. L. Latten, I. M. Shaw and B. E. Smith, *Fuel*, **64**, 509 (1985).
27. O. L. Gulder and B. Glavincevski, *Ind. Eng. Chem., Prod. Res. Dev.* **25**, 153 (1986).
28. J. F. Haw, T. E. Glass and H. C. Dorn, *Anal. Chem.*, **53**, 2332 (1981); *J. Magn. Reson.*, **49**, 22 (1982).
29. J. F. Haw, T. E. Glass, D. W. Hausler, E. Motell and H. C. Dorn, *Anal. Chem.*, **52**, 1135 (1980).
30. S. A. Knight, *Chem. Ind.* (*London*), 1920 (1967).
31. K. Saito, R. J. Battisberger, V. I. Stenberg and N. F. Woolsey, *Fuel*, **60**, 1039 (1981).
32. K. D. Bartle, A. Pomfret, A. J. Pappin, D. G. Mills and H. Evliye, *Org. Geochem.*, **11**, 139 (1987).
33. H. C. Dorn, L. T. Taylor and T. E. Glass, *Anal. Chem.*, **51**, 947 (1979).
34. C. E. Snape, W. R. Ladner and K. D. Bartle, *Anal. Chem.*, **51**, 2189 (1979).
35. Y. Maekawa, T. Yoshida and Y. Yoshida, *Fuel*, **58**, 864 (1979).
36. D. J. Cookson and B. E. Smith, *Fuel*, **62**, 34 (1983).
37. C. E. Snape, *Fuel*, **61**, 775 (1982); **62**, 621 (1983); in *Proceedings of the 1983 International Conference on Coal Science*, IEA, Pittsburgh, p. 624.
38. R. Gerhards, *Fresenius Z. Anal. Chem.*, **316**, 231 (1983); in L. Petrakis and J. P. Fraissard (eds), *Magnetic Resonance—Introduction, Advanced Topics and Applications to Fossil Energy*, Reidel, Dordrecht, 1984, p. 377.
39. P. F. Barron, M. R. Bendall, L. G. Armstrong and A. R. Atkins, *Fuel*, **63**, 1276 (1984).
40. L. W. Dennis and R. E. Pabst, paper presented at 24th Rocky Mountain Conference, Denver, 1982.
41. D. K. Dalling, G. Haider, R. J. Pugmire, J. Shabtai and W. E. Hull, *Fuel*, **63**, 525 (1984).
42. C. E. Snape and M. K. Marsh, *Am. Chem. Soc. Div. Pet. Chem., Prepr.*, **30**(2), 247 (1985).
43. D. J. Cookson and B. E. Smith, *Anal. Chem.*, **57**, 864 (1985).
44. J. N. Shoolery and W. L. Budde, *Anal. Chem.*, **48**, 1458 (1976).
45. M. Hajek, V. Skienar, G. Sebor, I. Lang and O. Weisser, *Anal. Chem.*, **50**, 773 (1978).
46. D. J. Cookson and B. E. Smith, *J. Magn. Reson.*, **57**, 355 (1984).
47. T. D. Alger, R. J. Pugmire, W. D. Hamill and D. M. Grant, *Am. Chem. Soc. Div. Fuel Chem., Prepr.*, **24**, 334 (1979).

48. V. Sklenar, M. Hajek, G. Sebor, I. Lang, M. Suckanek and Z. Starcuk, *Anal. Chem.*, **52**, 1794 (1980).
49. K. S. Seshadri, G. R. Raffaele, M. J. Douglas and H. P. Malone, *Fuel*, **57**, 549 (1978).
50. M. Aiura, T. Masunaga, K. Moriya and Y. Kageyama, *Fuel*, **63**, 1138 (1984).
51. C. E. Snape, W. R. Ladner and K. D. Bartle, *Fuel*, **64**, 1394 (1985).
52. J. R. Kershaw and R. I. Willing, *Liq. Fuels Technol.*, **2**, 33 (1984).
53. R. K. Harris and B. E. Mann (eds), *NMR of the Periodic Table*, Academic Press, New York, 1980.
54. D. W. Grandy, L. Petrakis, D. C. Young and B. C. Gates, *Nature (London)*, **308**, 175 (1984); in L. Petrakis and J. P. Fraissard (eds), *Magnetic Resonance—Introduction, Advanced Topics and Applications to Fossil Energy*, Reidel, Dordrecht, 1984, p. 689.
55. L. Cassidei, V. Fiandanese, G. Marchese and O. Sciacovelli, *Org. Magn. Reson.*, **22**, 486 (1984).
56. J. M. Novelli, M. Ngassoum, R. Faure, J. M. Ruil, L. Lena, E. J. Vincent and J. C. Escalier. In *Proceedings of International Symposium on the Characterisation of Heavy Crude Oils and Petroleum Residues*, Technip, Paris, 1984, p. 356.
57. D. D. McIntyre, O. P. Strausz and P. Otto, *Magn. Reson. Chem.*, **25**, 36 (1987).
58. C. E. Snape, C. A. Smith, K. D. Bartle and R. S. Matthews, *Anal. Chem.*, **54**, 20 (1982).
59. I. Schwager and T. F. Yen, *Anal. Chem.*, **51**, 569 (1979).
60. W. M. Coleman and A. R. Boyd, *Anal. Chem.*, **54**, 133 (1982).
61. K. D. Rose and C. G. Scouten, *AIP Conf. Proc.*, **70**, 82 (1981).
62. G. R. Leader, *Anal Chem.*, **45**, 1700 (1973).
63. K. D. Bartle, R. S. Matthews and J. W. Stadelhofer, *Appl. Spectros.*, **36**, 615 (1980).
64. K. D. Bartle, R. S. Matthews and J. W. Stadelhofer, *Fuel*, **60**, 1172 (1981).
65. P. S. Sleevi, T. E. Glass and H. C. Dorn, *Anal. Chem.*, **51**, 1931 (1979).
66. D. H. Finseth, G. J. Stiegel, C. E. Schmidt, R. F. Sprecher and R. G. Lett. In *Proceedings of the 1983 International Conference on Coal Science*, IEA, Pittsburgh, p. 180.
67. J. R. Havens, J. L. Koenig, D. Kuehn, C. Rhoads, A. Davis and P. C. Painter, *Fuel*, **62**, 936 (1983).
68. A. Pomfret, K. D. Bartle, S. Barrett, N. Taylor and J. W. Stadelhofer, *Erdöl Kohle, Erdgas, Petrochemie, Brennst.-Chem.*, **37**, 515 (1984).
69. K. D. Rose and M. A. Francisco, *Am. Chem. Soc. Div. Pet. Chem., Prepr.*, **30**(2), 762 (1985); *Energy Fuels*, **1**, 233 (1987).
70. M. A. Mikita, C. Steelink and R. L. Wershaw, *Anal. Chem.*, **53**, 1715 (1981).
71. R. Liotta, K. D. Rose and E. Hippo, *J. Org. Chem.*, **46**, 227 (1981).
72. J. R. Ratto, L. A. Heredy and R. P. Skowronski, *Am. Chem. Soc. Div. Fuel Chem Prepr.*, **24**, 155 (1979); *ACS Symp. Series*, No. 139, American Chemical Society, Washington DC, (1980) 370.
73. J. A. Franz, *Fuel*, **58**, 405 (1979).
74. D. C. Cronauer, R. I, McNeil, D. C. Young and R. G. Ruberto, *Fuel*, **61**, 610 (1982).
75. M. A. Wilson, P. J. Collin, P. F. Barron and A. M. Vassallo, *Fuel Process. Technol.*, **5**, 281 (1982).
76. S. A. Farnum, B. W. Farnum, J.R. Rindt, D. J. Miller and A. C. Wolfson, *Am. Chem. Soc. Div. Fuel Chem., Prepr.*, **29**(1), 144 (1984).
77. P. J. Collin and M. A. Wilson, *Fuel*, **62**, 1243 (1983).
78. J. M. Dereppe, J. P. Boudou, C. Moreaux and B. Durand, *Fuel*, **62**, 575 (1983).
79. T. Yoshida, Y. Nakata, R. Yoshida, S. Ueda, N. Kanda and Y. Maekawa, *Fuel*, **61**, 824 (1982).
80. M. J. Sullivan and G. E. Maciel, *Anal. Chem.*, **54**, 1606 and 1615 (1982).

81. L. B. Alemany, D. M. Grant, R. J. Pugmire and L. M. Stock, *Fuel*, **63**, 513 (1984).
82. S. A. Farnum, D. D. Messick and B. W. Farnum, *Am. Chem. Soc. Div. Fuel Chem.*, *Prepr.*, **31**(1), 60 (1986).
83. D. L. Vander Hart and H. L. Retcofsky, *Fuel*, **55**, 202 (1976).
84. W. T. Dixon, J. Schaefer, M. D. Sefcik, E. O. Stejskal and R. A. McKay, *J. Magn. Reson.*, **49**, 341 (1982).
85. D. E. Axelson, *Fuel*, **66**, 197 (1987).
86. R. A. Wind, M. J. Duijvestijn, C. van der Lugt, J. Smidt and J. Vriend. In L. Petrakis and J. P. Fraissard (eds), *Magnetic Resonance—Introduction, Advanced Topics and Applications to Fossil Energy*, Reidel, Dordrecht, 1984, p. 461.
87. R. A. Wind, *Am. Chem. Soc. Div. Fuel Chem.*, *Prepr.*, **31**(1), 223 (1986).
88. M. A. Wilson, R. J. Pugmire, J. Karas, L. B. Alemany, W. R. Woolfenden, D. M. Grant and P. H. Given, *Anal. Chem.*, **56**, 933 (1984).
89. K. D. Schmitt and E. W. Sheppard, *Fuel*, **63**, 1241 (1984).
90. E. W. Hagaman and M. C. Woody in *Proceedings of the 1981 Int. Conference on Coal Science*. Verlag Glückauf, Essen (1981), p. 807.
91. P. D. Murphy, T. J. Cassidy and B. C. Gerstein, *Fuel*, **61**, 1233 (1982).
92. R. J. Pugmire, W. R. Woolfenden, C. L. Mayne, J. Karas and D. M. Grant, *Am. Chem. Soc. Div. Fuel Chem.*, *Prepr.*, **28**(1), 103 (1983).
93. T. Terao, H. Miura and A. Saika, *J. Am. Chem. Soc.*, **104**, 5228 (1982).
94. M. I. Burgar, J. R. Kalman and J. F. Stephens. In *Proceedings of the 1985 International Conference on Coal Science*, Pergamon Press, Sydney, 1985, p. 780.
95. S. J. Davenport and R. H. Newman. In *Proceedings of the 1985 International Conference on Coal Science*, Pergamon Press, Sydney, 1985, p. 784.
96. M. J. Trewhella, I. J. F. Poplett and A. Grint, *Fuel*, **65**, 541 (1986).
97. L. B. Alemany, D. M. Grant, R. J. Pugmire, T. D. Alger and K. W. Zilm, *J. Am. Chem. Soc.*, **105**, 2133 and 2142 (1983).
98. B. C. Gerstein. In C. Karr (ed), *Analytical Methods for Coal and Coal Products*, Vol. III, Academic Press, New York, 1979, p. 425.
99. K. J. Packer, R. K. Harris, A. M. Kenwright and C. E. Snape, *Fuel*, **62**, 999 (1983).
100. R. E. Botto, R. Wilson, R. Hayatsu, R. L. McBeth, R. G. Scott and R. E. Winans, *Am. Chem. Soc. Div. Fuel Chem.*, *Prepr.*, **30**(4), 180 (1985).
101. E. W. Hagaman, R. R. Chambers, Jr, and M. C. Woody, *Anal. Chem.*, **58**, 387 (1986).
102. R. L. Dudley and C. A. Fyfe, *Fuel*, **61**, 651 (1982).
103. R. Gerhards and I. Kasueschke. In *Proceedings of the 1985 International Conference on Coal Science*, Pergamon Press, Sydney, 1985, p. 838.
104. F. P. Miknis, G. E. Maciel and V. J. Bartuska, *Org. Geochem.*, **1**, 169 (1979).
105. F. P. Miknis and G. E. Maciel. In L. Petrakis and J. P. Fraissard (eds), *Magnetic Resonance-Introduction, Advanced Topics and Applications to Fossil Energy*, Reidel, Dordrecht, 1984, p. 545.
106. P. G. Hatcher, I. A. Breger, N. M. Szeverenyl and G. E. Maciel, *Org. Geochem.*, **4**, 9 (1982).
107. K. W. Zilm, R. J. Pugmire, S. R. Larter, J. Allen and D. M. Grant, *Fuel*, **60**, (1981).
108. M. A. Wilson, R. J. Pugmire, A. M. Vassallo, D. M. Grant, P. J. Collin and K. W. Zilm, *Ind. Eng. Chem. Prod. Res. Dev.*, **21**, 477 (1982).
109. D. H. Finseth, D. L. Cillo, R. F. Sprecher, H. L. Retsofsky and R. G. Lett, *Fuel*, **64**, 1718 (1985).
110. L. B. Alemany and L. M. Stock, *Fuel*, **61**, 1088 (1982).
111. R. E. Botto, R. Hayatsu, R. G. Scott, R. L. McBeth and R. E. Winans. In *Proceedings of the 1985 International Conference on Coal Science*, Pergamon Press, Sydney, (1985), p. 624.

112. D. P. Burum and W. K. Rhim, *J. Chem Phys.*, **71** 944, (1979).
113. H. Rosenberger, G. Scheler and K.-H. Rentrop, *Z. Chem.*, **23**, 34 (1983).
114. K. W. Zilm and G. G. Webb, *Am. Chem. Soc. Div. Fuel Chem., Prepr.*, **31**(1), 230 (1986).
115. P. T. Ford, N. J. Friswell and I. J. Richmond, *Anal. Chem.*, **49**, 594 (1977).
116. S. D. Robertson, F. Cunliffe, C. S. Fowler and I. J. Richmond, *Fuel*, **58**, 771 (1979).
117. A. Jurkiewicz, A. Marzec and S. Idziak, *Fuel*, **60**, 1167 (1981).
118. P. H. Given, A. Marzec, W. A. Barton, L. J. Lynch and B. C. Gerstein, *Fuel*, **65**, 155 (1986).
119. A. Jurkiewicz, A. Marzec and N. Pislewski, *Fuel*, **61**, 647 (1982).
120. W. A. Barton, L. J. Lynch and D. S. Webster, *Fuel*, **63**, 1262 (1984).
121. L. J. Lynch and D. S. Webster. In *Proceedings of the 1983 International Conference on Coal Science*, IEA, Pittsburgh, 1983, p. 663.
122. K. Miyazawa, T. Yokono and Y. Sanada, Carbon, **17**, 223 (1979).
123. E. M. Dickinson, *Fuel*, **59**, 290 (1980).
124. D. M. Cantor, *Anal. Chem.*, **50**, 1185 (1978).
125. S. Yokoyama, H. Uchino, T. Katoh, Y. Sanada and T. Yoshida, *Fuel*, **60**, 254 (1981).
126. C. E. Snape, W. R. Ladner, L. Petrakis and B. C. Gates, *Fuel Process. Technol.*, **8**, 155 (1984).
127. D. T. Allen, L. Petrakis, D. W. Grandy, G. R. Gavalas and B. C. Gates, *Fuel*, **63**, 803 (1984).
128. P. S. Shenkin, *Am. Chem. Soc. Div. Pet. Chem., Prepr.*, **28**(5), 1367 (1983).
129. C. E. Snape and K. D. Bartle, Fuel, **63**, 883 (1984).
130. K. D. Bartle, C. Gibson, M. J. Mulligan, N. Taylor, T. G. Martin and C. E. Snape, *Anal. Chem.*, **54**, 1730, (1982).
131. B. C. Bockrath and F. K. Schweighardt. In J. W. Bunger and N. C. Li (eds), *Chemistry of Asphaltenes*, Advances in Chemistry Series, No. 195, American Chemical Society, Washington, DC. 1981, p. 29.
132. M. Farcasiu, *Am. Chem. Soc. Div. Fuel Chem., Prepr.*, **24**(1), 121 (1979).
133. B. C. Bockrath and R. P. Noceti, *Fuel Process. Technol.*, **2**, 143 (1979).
134. J. R. Kershaw, *Liquid. Fuels Technol.*, **3**, 205 (1985).
135. S. Yokoyama, D. M. Bodily and W. J. Wiser, *Fuel*, **58**, 162 (1979); *Am. Chem. Soc. Div. Fuel Chem., Prepr.*, **26**(2), 38 (1981).
136. L. Schanne and M. H. Haenel, *Fuel*, **60**, 556 (1981).
137. D. G. Richards, C. E. Snape, K. D. Bartle, C. Gibson, M. J. Mulligan and N. Taylor, *Fuel*, **62**, 724 (1983).
138. T. T. Le and D. T. Allen, *Fuel*, **64**, 1754, (1985).
139. D. T. Allen, *Am. Chem. Soc. Div. Pet. Chem., Prepr.*, **30**, 270 (1985).
140. E. W. Hagaman, R. R. Chambers, Jr, and R. L. Rardin. In *Proceedings of the 1985 International Conference on Coal Science*, Pergamon Press, Sydney, 1985, p. 778.

Analytical NMR
Edited by L. D. Field and S. Sternhell
© 1989 John Wiley & Sons Ltd

Chapter 5

NMR of Zeolites, Silicates and Solid Catalysts

A. D. H. Clague

Thornton Research Centre, Shell Research Ltd, PO Box 1, Chester CH1 3SE, UK

N. C. M. Alma

Koninklijke/Shell Laboratorium Amsterdam, Badhuisweg 3, 1031 CM Amsterdam, The Netherlands

1 INTRODUCTION

The analytical potential of NMR in studies of zeolites, silicates and catalysts has made dramatic progress in recent years. Important contributory factors have been the improved multinuclear capabilities, improved sensitivities of NMR spectrometers and the ability to study solids in a routine manner.

In this chapter we aim to describe the variety of ways in which analytical NMR can be applied in the investigation of solid (carrier) materials used in heterogeneous catalysis. General requirements for these types of materials are that they must be stable under the conditions of the catalytic reaction (often elevated temperatures and pressures) and that they should have a high surface area accessible to substrates and reactants. In many cases a metal (or several metals) in ionic or neutral form is attached to the surface.

Zeolites are a class of materials that fulfil these requirements. Basically, zeolites are porous, crystalline aluminosilicates, having a three-dimensional framework structure of SiO_4 and AlO_4 tetrahedra, where every tetrahedral silicon or aluminium is attached via oxygen bridges to four other tetrahedra; according to the Loewenstein rule[1], no direct Al—O—Al bonds between tetrahedral aluminium atoms are allowed. Other trivalent ions, (e.g. Ga, B, Fe) may also be substituted in place of aluminium. Without incorporated aluminium (or other trivalent ions) the framework is neutral, but otherwise an excess negative charge of the framework arises and this is compensated either by protons, associated with a lattice oxygen, or by other cations. The protons, forming Brönsted acid sites, or the Lewis acid sites generated on dehydroxylation may also play an important role in catalytic processes.

Aluminosilicate zeolites are naturally occurring minerals, but synthetic zeolites are commonly used as catalysts. The stability of these materials is increased by the so-called ultrastabilization process, which was shown to be associated with removal of aluminium from the framework. Aluminium is replaced by silicon from elsewhere in the zeolite.

At least 40 types of zeolites are known, with different lattice structures, having one-, two- or three-dimensional pores, with diameters varying between 2.5 and 7Å. Framework structures of zeolites have been elucidated by X-ray diffraction (XRD). Unfortunately, because of the nearly equal scattering powers of silicon and aluminium, no discrimination can be made between these two nuclei, so that the aluminium locations could not be established. Among the best known and most intensively studied zeolite structures are zeolite A (Figure 1), the faujasite-type zeolites Y and X (Figure 2), mordenite (Figure 3) and zeolite ZSM-5 (Figure 4).

The exact pore size and shapes are of great interest since these are responsible for another attractive property of zeolitic catalysts (carriers), namely their shape selectivity. For example, the three-dimensional pore system of zeolite A cannot accommodate aromatic molecules with six-membered rings, whereas the faujasite (three-dimensional), mordenite (one-dimensional) and ZSM-5 (two-dimensional) zeolites allow the entry of such aromatic molecules with different degrees of substitution[2]. Extensive literature is available on zeolite structures and their synthesis[2-7].

Recently, another class of microporous crystalline materials have been synthesized, namely the crystalline aluminophosphates[8], consisting of alternating linked AlO_4 and PO_4 tetrahedra of which a variety of structural modifications are now available, some of which are structurally identical with known

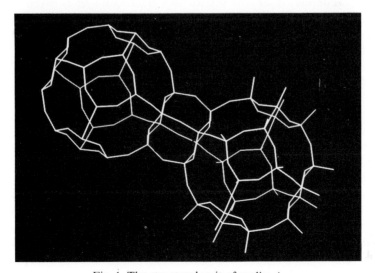

Fig. 1. The structural unit of zeolite A

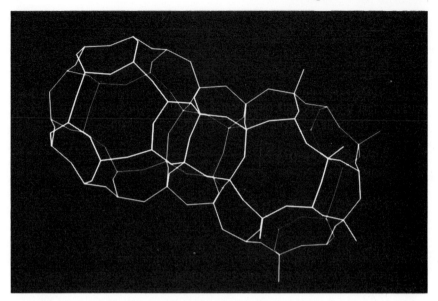

Fig. 2. The structural unit of faujasite

aluminosilicate strutures. In these structures substitution of either of the tetrahedral sites may be accomplished (B, Si, metals), generating a charged framework structure.

Other materials which have application in catalysis are the clays, which are built up from stacked two- or three-layer sheets; the sheets consist of an octahedral (alumina) layer and one or two tetrahedral silica layers. A large variety of substitutions in both types of layers may occur. A typical layer structure of a dioctahedral 2:1 layer silicate clay is depicted in Figure 5. The general formula of this structure is given by:

$$\{n H_2 O, (z/x) M^{z+}\}_{interlayer} \cdot (Si_{4-x}, Al_x)_{tetrahedral} \cdot O_{10}(OH)_2 \cdot (Al_2)_{octahedral} \quad (1)$$

The M^+ ions in the interlayers provide charge balance. The surface area available for reactants depends on the interlayer spacing. To obtain a structure with a stable high interlayer spacing, pillars of alumina can be built between the layers. The catalytic properties are again strongly dependent on the degree and site of substitution.

Finally, there are materials with less ordered structures that are important as catalyst carriers. These are amorphous silica, several structural modifications of alumina (with different degrees of hydration) and amorphous aluminosilicates. These materials can be prepared with different surface areas. Surface areas and pore volumes are commonly determined via nitrogen sorption techniques.

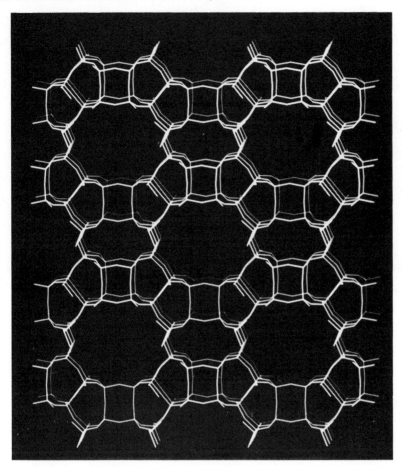

Fig. 3. The structure of mordenite

NMR has found application in characterizing overall and surface structural properties.

Following the early (1980) extensive studies of Lippmaa and co-workers[9,10] on minerals and zeolites, NMR has now established itself as a major and unique analytical tool in the characterization of the structural and mobility properties of zeolites and other (alumino) silicate materials used in hetergeneous catalysis. A considerable number of research groups have contributed to the extensive literature now available in this area and every day new exciting data are produced. In this light it is a virtually impossible task to survey the whole area, and to give justice to all valuable investigations that have been carried out. Fortunately, a number of good reviews are available, highlighting the accomplishments in specific areas. In particular we mention the papers by Klinowski,

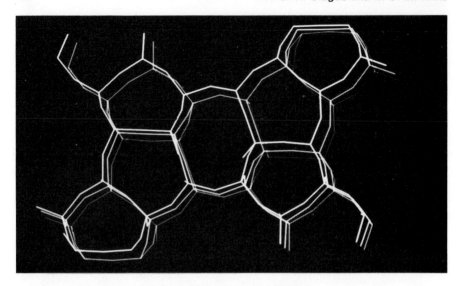

Fig. 4. The structure of ZMS-5

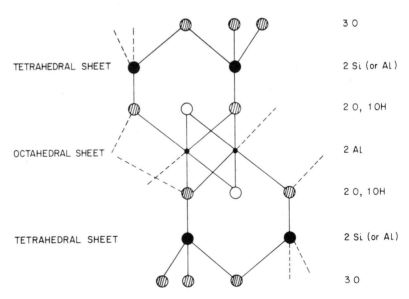

Fig. 5. Schematic diagram of the layer structure of a dioctahedral three-layer clay

Fyfe, Nagy and co-workers dealing the NMR of zeolites[11–15] and a recent review by Austermann et al.[16] dealing with catalyst characterization. We shall refer to these reviews whenever appropriate, and highlight some of the most recent developments. Our aim is that by the end of the chapter the reader will have some sense of the analytical capabilities of NMR and the challenges that it is facing in this field in research.

The bulk of this chapter will focus closely on the NMR study of solids without, however, neglecting the use of liquid-state NMR, particularly in the synthesis of silicates and zeolites.

2 NMR OF SOLIDS

Until the early 1970s, analytical applications of NMR were restricted to studies of the liquid state. This was due, of course, to the fact that the production of a high-resolution spectrum of a solid was considerably more difficult than that of a liquid. Factors which contribute to this stem mainly from the lack of molecular mobility within a solid, resulting in line broadening and potentially long relaxation times. That these problems have been largely overcome is evidenced by the many recent publications[17–19] dealing with high-resolution spectra of solids, including the area of zeolites, silicates and solid catalysts. Since an understanding of the solid-state NMR technique is of critical importance to the interpretation of analytical results, we shall devote this section to a discussion of the special characteristics of solid-state NMR.

2.1 Linewidths

It is of interest to dwell briefly on how the lack of molecule motion in solids leads to line broadening, which, from an analytical viewpoint, is highly undesirable because if lines are overlapping the important information available from the isotropic chemical shifts becomes irretrievable. There are three principle line-broadening mechanisms, outlined below.

2.1.1 Dipolar interactions

In addition to the applied magnetic field experienced by a magnetic nucleus, there may be additional magnetic interactions with the near environment. One of the most important of these is a dipolar interaction with other magnetic nuclei, the strength of which depends on the magnitude of the neighbouring dipole, the distance and the orientation of the internuclear vector with respect to the external field. (See Ref. 17, p. 164, for a summary of the theory). In a liquid system, rapid molecular motion averages the dipolar interactions to zero; in a solid, however, no such averaging occurs, and a dipolar-broadened signal is obtained. Unless one is dealing with a small isolated group of spins, the shape of the line is broad and featureless, and $\Delta v_{1/2}$ may vary from a few Hertz to more than 20 kHz. There are two forms of dipolar interaction which lead to line broadening: (a) homonuclear, in which the interaction occurs between the spins of like nuclei, and (b) heteronuclear, in which the interaction occurs between the nuclear species under observation and spins of different nuclei. From a catalytic viewpoint, the most important interactions are between ^{13}C and ^{1}H and between ^{29}Si and ^{1}H. Because of the low natural abundance of ^{13}C and ^{29}Si and the strong distance dependence of dipolar interactions, homonuclear interaction usually does not play a significant role for these nuclei.

2.1.2   Chemical shift anisotropy (CSA)

Details of chemical shift theory are dealt with in several standard texts[20-23]. The chemical shift is a tensor quantity with three principal components σ_{11}, σ_{22} and σ_{33}, the values of which normally differ. The observed chemical shift, for a given nucleus, depends on the orientation of the molecule and of the chemical bond containing the nucleus, relative to the external magnetic field. In a polycrystalline powder, this chemical bond will have a distribution of orientations relative to the applied field, resulting in a distribution of chemical shifts. The spectrum will have the appearance of a powder pattern. The theoretical shape for a non-axially symmetric chemical shift tensor is shown in Figure 6.

The isotropic shift, σ, containing important chemical information is given by

$$\sigma = \tfrac{1}{3}(\sigma_{11} + \sigma_{22} + \sigma_{33}) \tag{2}$$

The isotropic shift can only be derived from this pattern by lineshape simulation if the signal does not overlap with signals of other species. In such favourable cases a knowledge of the full tensor yields additional chemical information. Typically, however, the resolution of several species is required, and techniques to remove CSA are generally utilized (two-dimensional experiments are now possible, which yield both isotropic chemical shift and information on the chemical shift tensor).

2.1.3   Linewidths and intensities of quadrupolar nuclei

Since some of the major nuclei considered in this chapter, ^{27}Al ($I = \tfrac{5}{2}$), ^{17}O ($I = \tfrac{5}{2}$) and ^{11}B ($I = \tfrac{3}{2}$), are quadrupolar, we need to consider briefly the difficulties (and also the special opportunities) associated with the investigation of these nuclei with half-integer spins. In addition to their magnetic moment, these nuclei have an electric quadrupole moment of magnitude Q, which is the

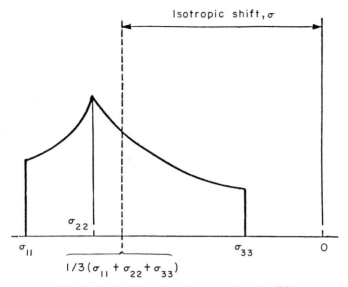

Fig. 6. Typical lineshape for a powdered solid

result of the non-spherical distribution of nuclear charge. When placed in an electric field gradient, the interaction between this gradient and the quadrupole moment (the so-called quadrupolar interaction, QI) yields an additional orientation-dependent contribution to the nuclear energy. The electric field gradient (V) is a tensor quantity defined in a molecular axis system in which it is diagonal with $V_{zz} > V_{xx} > V_{yy}$, where the deviation from cylindrical symmetry η is given by

$$\eta = (V_{xx} - V_{yy})/V_{zz} \tag{3}$$

For perfect octahedral or tetrahedral symmetry the electric field is zero. In practice, however, small distortions are present, resulting in considerable QI. Explicit expressions for the orientation-dependent changes in energy levels and transition frequencies are available[24-31]. The major conclusions from the extensive theoretical investigations are as follows:

(a) the central $(\frac{1}{2}, -\frac{1}{2})$ transition is unaffected by quadrupolar interaction to first order. The other transitions are broadened by an amount given approximately by v_Q where

$$v_Q = (e^2qQ/h)\{3/[2I(2I - 1)]\} \tag{4}$$

where e^2qQ/h is known as the quadrupolar coupling constant (QCC) with $eq = V_{zz}$. For aluminium in aluminosilicates, typical values of QCC are

1–9 MHz. The first-order QI can be averaged out by magic angle spinning (MAS) (Section 2.2.1), resulting in the generation of numerous sidebands over a frequency range of the order of v_Q, for the non-central transitions.

(b) To second order, the QI broadens all transitions by an amount proportional to v_q^2/v_{Larmor}; this broadening is much smaller than the first-order broadening and is *inversely proportional* to the magnetic field. It gives rise to a complicated lineshape, depending on η and the QCC[24–29,32,33]. MAS cannot completely average out this broadening, but lines are narrowed by a factor of ca 4. The centre of gravity of all transitions (with and without MAS) is shifted from the isotropic chemical shift by an amount proportional to v_q^2/v_{Larmor}, which is different for all transitions[24,25,30].

As a consequence, only the $(\frac{1}{2}, -\frac{1}{2})$ transition is observed in the normal MAS NMR experiment, sometimes together with sidebands of the other transitions. Other transitions are either not or incompletely exited. Therefore, the intensity of a resonance is less than that in a liquid-state experiment, where all transitions contribute to the observed signal, and also the limited bandwidth of excitation results in a complex dependence of signal intensity and shape on the pulse length[32,34,35] (see Figure 7.) To be able to use signal intensities quantitatively,

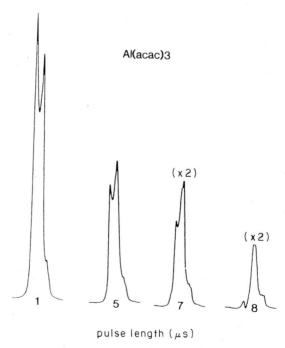

Fig. 7. Illustration of the complex dependence of the ^{27}Al lineshape and intensity on the length of the excitation pulse, for solid Al(acetylacetone)$_3$

one should firstly verify that indeed, for all species compared, only the $(\frac{1}{2}, -\frac{1}{2})$ transition is observed (or use a correction factor) and secondly use only short excitation pulses[35] with $(I + \frac{1}{2})\omega_{r.f.}t_{pulse} \leqslant \pi/6$. The observed chemical shift is not equal to the isotropic chemical shift and depends on the magnetic field strength: the higher the magnetic field, the narrower is the line and the more the shift approaches the isotropic shift.

A method that makes use of the complex pulse length dependence is the so-called nutation NMR experiment. This is essentially a simple two-dimensional method, where magnetization is allowed to evolve during an incremented period t_1, under the influence of a strong r.f. field[32,35,36]. This method allows discrimination between species of similar shift but different QCC.

2.2 Line narrowing procedures

2.2.1 Magic angle spinning

It has already been mentioned that rapid molecular motion, as in liquids, causes line narrowing by averaging out the dipolar broadening and chemical shift anisotropy (see Chapter 2, Section 2.1). Magic angle spinning (MAS) is a technique which, in effect, provides a substitute for this molecular motion in solids and thereby allows the direct observation of the isotropic chemical shift. The technique has been recognized for ca 30 years[37-39] and more recent treatises deal with the quantitative aspects of the theory[20,22,40].

Even slow MAS results in a dramatic narrowing of lines broadened by CSA and first-order QI; in this case narrowed lines are generated with sidebands occurring at integral multiples of the spinning rate. The range of chemical shift anisotropies varies considerably with the nucleus under study, and typical values for the common nuclei are tabulated in Ref. 17, p. 169. Broadening by dipolar coupling can be fully eliminated only if the spinning speeds are higher than the strengths of dipolar coupling (usually ca 20 kHz). Also in the case of CSA, one would choose to avoid sidebands by fast spinning. Since the anisotropy (in Hertz) is proportional to the magnetic field strength, higher spinning speeds need to be used at higher fields. An extreme example for a nucleus with a large anisotropy is given in Figure 8a. With a spin rate which is too low, the resulting spectrum is highly confusing (Figure 8b). Increasing the spinning speed improves the situation (Figure 8c), but does not eliminate the sideband problem. If, however, the spinning speed is the maximum achievable, overlap of sidebands with other lines and assignment may become an important complication. In such cases the isotropic chemical shift may be identified by varying the spinning speed, whereby only the centreband will remain invariant, or by making use of sideband suppression procedures[41-43]. Although the latter

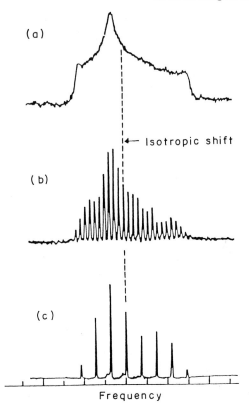

Fig. 8. Effect of spinning rate on the spectrum of a highly anisotropic nucleus: (a) static sample; (b) slow spinning; (c) faster spinning

approach would appear to be a panacea, it suffers from lack of quantitativeness. In extreme cases one might consider using a lower field.

2.2.2 Variable angle spinning

When dealing with quadrupolar nuclei, spinning at the magic angle does not necessarily lead to the narrowest lines attainable. A modification, referred to as variable angle spinning (VAS), has been introduced[33,44,45], which takes account of the fact that the dominant line-broadening mechanism may well arise from second-order QI. The angle of rotation for optimum line narrowing depends in this case on the asymmetry parameter η (equation 3), as a result of which the resonance may be shifted far from its isotropic chemical shift, depending on QCC. In some cases this feature may be exploited for resolving lines of similar shift but different QCC. Rotation at an angle other than the magic angle, however, fails to remove CSA, in which case one may be worse off in terms of line narrowing. As mentioned earlier, a compromise must be sought, which may in fact require the use of a different magnetic field strength.

2.2.3 Homonuclear dipolar decoupling

Modern commercial spectrometers allow the operator to apply sequences of carefully tailored pulse cycles to a sample and these can eliminate or partially remove the effects of the homonuclear interaction. These procedures have provided a wealth of alternative, so-called 'multiple pulse', sequences[17,20,22,46].

All of the multiple-pulse techniques are in practice complicated and there are many pitfalls. It is therefore recommended that potential experimentalists refer to the literature[20,47-50] and to the summary in Ref. 17, p. 196.

2.2.4 Heteronuclear dipolar decoupling

The physical concept of heteronuclear decoupling is more familiar than that of homonuclear decoupling. It is in fact no more than an extension of the normal proton decoupling technique routinely used in all liquid-state ^{13}C NMR, the object being to average to zero the interactions of the abundant spins, designated **I** (usually ^{1}H), with the rare spins under observations, designated **S**, e.g. ^{13}C, ^{29}Si, ^{15}N. Decoupling is achieved simply by applying radiofrequency (r.f.) power at the Larmor frequency of the **I** spins, which causes rapid spin flipping of these spins with the result that there is no effect of the **I** spins on the **S** spins. Typically, r.f. field strengths of the order of 10 G are required for effective decoupling in solids.

2.2.5 Zero-field NMR

Although not specifically a line-narrowing technique, zero-field NMR is a new and promising method which is particularly suited to the study of quadrupolar nuclei, because it avoids orientation-dependent broadening. In this field cycling technique, the evolution of the magnetization at zero field is measured with the sensitivity of high-field NMR. Recent descriptions of the principles and the potentials of the techniques are given in Refs 51 and 52. An important limitation of this method is that so far it is applicable only to materials with reasonably long T_1.

2.3 Relaxation times

The T_1 relaxation times in solids may be extremely long, whereas T_2 is usually short, and one should check this parameter carefully if quantitative data are required. For ^{29}Si the T_1 values have been found to vary between 4 and 5000 s[53]. For highly siliceous zeolites, it was demonstrated that molecular oxygen may be the dominant relaxing agent: recording spectra under O_2 reduced the T_1 relaxation time from 230 to 1.35 s[54]. Useful discussions of relaxation in solids were given by Abragam[23] and Goldman[55].

A relaxation parameter which, apart from T_1 and T_2, is useful for analytical purposes is that of relaxation in the rotating frame, $T_{1\rho}$, it resembles T_1, except that it refers to the rate of change of transverse magnetization along the direction of an on-resonance spin-locking magnetic field $\mathbf{B}_{eff}$. Since $\mathbf{B}_{eff}$ is usually many orders of magnitude less than $\mathbf{B}_0$, it enables one to probe molecular motion effects on a very different time scale than with T_1 studies alone. Relaxation-time studies are finding increasing use in characterizing catalytically important species and for investigating molecular motion at a surface[17]. Another important relaxation parameter is T_{1CP}, which characterizes the rate of magnetization transfer in the presence of matched spin-locking fields for two dipolar coupled spin systems $\mathbf{I}$ and $\mathbf{S}$. This parameter will be discussed in the next section. A theoretical treatment of relaxation phenomena in solids was given by Mehring[20] and Spiess[56].

2.4 Cross-polarization

Cross-polarization allows transfer of magnetization from an abundant species, $\mathbf{I}$, to a rare species, $\mathbf{S}$. The benefits are primarily an intensity enhancement of the $\mathbf{S}$ signal by a factor of γ_I/γ_S (i.e. a factor 5 for ^1H–^{29}Si) and a reduction of the recycle time between experiments since the rate-determining relaxation time is now that of the $\mathbf{I}$ species, rather than of the $\mathbf{S}$ spins ($\mathbf{I}$ relaxation is usually much faster than $\mathbf{S}$ relaxation). The phenomenon relies on bringing $\mathbf{I}$ and $\mathbf{S}$ spins in states with equal fluctuation frequency of the z-magnetization, whereby mutual spin flips of $\mathbf{I}$ and $\mathbf{S}$ spins are possible. This is accomplished by simultaneously applying 'matching' r.f. fields to the $\mathbf{I}$ and $\mathbf{S}$ spins where the Hartmann–Hahn[57] matching condition ($\gamma_I \mathbf{B}_{1,I} = \gamma_S \mathbf{B}_{1,S}$) is fulfilled.

One may make use of differences in magnetization transfer rates (characterized by T_{1CP}) for discrimination between chemical species. Rigid or immobilized species experience a more efficient magnetization transfer process than more mobile molecules. Differences of a factor of 10 in T_{1CP} can readily occur. In addition to providing motional information, there is another benefit to be derived from CP studies and this depends on the fact that the effectiveness of the CP phenomenon is proportional to r^{-3}, where r is the distance between the $\mathbf{I}$ and the $\mathbf{S}$ nuclei. The efficiency of transfer of magnetization falls off extremely rapidly as the distance between $\mathbf{I}$ and $\mathbf{S}$ increases. Bulk Si nuclei in SiO_2, for example, will be relatively distant from any proton species, whereas Si nuclei at the surface will be relatively close to surface OH groups (or any tightly bound organic species). The use of cross-polarization ^{29}Si NMR will therefore enable one to examine Si nuclei at the surface separately from those in the bulk material.

Since the CP process relies on the availability of dipolar couplings, a problem may be encountered when it is applied at high fields. If high MAS rates are to be used, dipolar coupling may be fully averaged, thereby destroying the CP efficiency.

2.5 Solid-state instrumentation

The foregoing sections have provided an indication of the additional complexities involved in the study of solids. There are therefore some additional demands placed on the NMR instrumentation to allow one to produce satisfactory spectra. The most important points will be dealt with here while providing reference material to more detailed texts.

2.5.1 High-field magnets

Strong magnetic fields (i.e. greater than 2.5 T) are achievable only with superconducting magnets. Although the lower-field superconducting magnets are not substantially more expensive than electromagnets, for the higher fields (500 or 600 MHz ^{1}H frequency) expense is still a major obstacle. The main point of concern with higher magnetic fields is that higher spin rates are necessary to average out the CSA. Although until recently maximum routine spinning rates were ca 5 kHz, instrumentation offering spinning rates over 9 kHz is now becoming commercially available, while spin rates in access of 20 kHz have already been reported. The benefits of high fields lie in three principle factors: (a) the sensitivity is roughly proportional to the magnetic field to the power $\frac{3}{2}$; small samples or adsorbed species therefore benefit enormously from this extra sensitivity; (b) for quadrupolar nuclei with half-integar spin, lines tend to become narrower at higher fields, since the second-order quadrupolar broadening is inversely proportional to the magnetic field; and (c) since the chemical-shift dispersion is proportional to the field, potentially more chemical-shift information becomes available at higher field strengths. This has been well demonstrated by Fyfe et al.[58] for a number of nuclei.

2.5.2 Magic angle spinning facilities

Magic-angle spinning is an essential requirement for the high-resolution NMR of solids, as (apart from zero-field NMR) it is the only means for removing the effects of chemical shift anistropy. The theory of rotor design and high-speed spinning is well established[59]. There are a number of systems available from various instrument manufacturers and all tend towards double-bearing rotors and away from the traditional 'mushroom' type of rotor pioneered by Andrew et al.[37]. An important advantage of double-bearing systems is that they can be easily operated at variable temperatures. Pneumatic loading systems are also available and these improve the ease of operation. Nevertheless, high-speed spinning is prone to failure, and strict guarantees should be obtained on this point before purchase. Rotor materials vary, depending on the nucleus under study. For ^{13}C, a non-carbon-containing material is most desirable and armoured boron nitride[60] has been used. For most nuclei other than ^{13}C and ^{1}H good results have been obtained with Delrin and Vespel. Alumina is now very frequently used and Kel-F has been used for ^{1}H measurements. Ceramics such as macor, SiN_3 and zirconia (although the first two cannot be used for ^{29}Si NMR) initially suffered from abrasive qualities, but as experience grows they are becoming increasingly used as rotor materials, particularly as variable-temperature experiments become commonplace in the study of catalysts.

2.5.3 Electronics

The following spectrometer features are needed in order to be able to perform the bulk of solid-state experiments:

(a) High-power radiofrequency electronics. The system must produce and transmit r.f. power levels of several hundred watts, and the probe circuit must not arc during the (prolonged) high-power pulses. Pulse timings, widths, amplitudes and phases must be accurately adjustable, stable and flexibly programmed; dead times should be minimized. Cross-polarization facilities are essential. Fast ADCs are necessary for wide-line applications.
(b) Multinuclear capability. This hardly needs further justification if one intends to study systems of catalytic importance. The basic frequency range should be designed to encompass the nuclei ^{1}H, ^{2}H, ^{7}Li, ^{11}B, ^{13}C, ^{15}N, ^{19}F, ^{23}Na, ^{27}Al, ^{29}Si, ^{31}P, ^{51}V, ^{55}Mn, ^{59}Co, ^{119}Sn, ^{129}Xe, ^{133}Cs, ^{195}Pt and ^{205}Tl. The key nuclei having analytical potential are ^{29}Si, ^{13}C, ^{27}Al and ^{1}H and the bulk of the following text will involve these nuclei.

3   STRUCTURAL INFORMATION

3.1   Lattice structures

3.1.1   Characterization of zeolite frameworks

(a) $^{29}Si\ NMR$

It is only thanks to the capabilities of the X-ray diffraction (XRD) techniques that we now know the wide variety of structures in naturally occurring and synthetic zeolites. Recently, high-resolution electron microscopy (HREM) has also contributed considerably to our knowledge of these materials, particularly for those structures which are only accessible with difficulty by XRD analysis owing to the lack of reasonably sized crystals. Whereas XRD and HREM are valuable for the determination of long-range order, NMR can be used to probe the immediate environments of the nuclei studied, and may therefore yield complementary information. An important result of Lippmaa et al.'s work[9] was the establishment of the sensitivity of the ^{29}Si chemical shift to the degree of condensation of the silicon–oxygen tetrahedra. To characterize the degree of condensation, a SiO_4 unit is symbolised by Q^n ($n = 0, 1, 2, 3$ or 4), where n is the number of Si or Al tetrahedra connected to the SiO_4 unit in question. Chemical shift ranges for the different Q^n units are given in Figure 9. For zeolites, a very useful feature is that the NMR chemical shift of the ^{28}Si resonance is sensitive to aluminium substitution in a neighbouring tetrahedral site. As demonstrated by Lippmaa et al.[10], the silicon resonance generally shifts to low field by ca 5 ppm for every additional aluminium. Therefore five (somewhat overlapping) shift ranges may be discriminated in the ^{29}Si NMR spectrum of a zeolite. These belong to Si(nAl) with $n = 0, 1, 2, 3,$ or 4, where n indicates the number of aluminium tetrahedra sharing oxygens with the SiO_4 tetrahedron under consideration (see Figure 9).

From the intensities of the resonances found in these different regions, the silicon to aluminium ratio may be calculated[12] using

$$(Si/Al)_{NMR} = \sum_{n=0}^{n=4} I_{Si(nAl)} \bigg/ \sum_{n=0}^{n=4} 0.25n I_{Si(nAl)} \qquad (5)$$

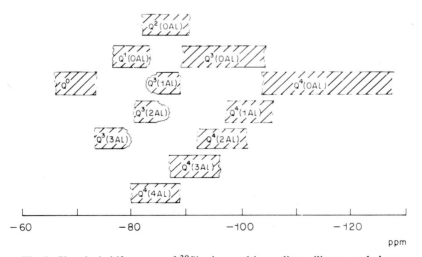

Fig. 9. Chemical shift ranges of ^{29}Si, observed in zeolites, silicates and clays

This has proved to be extremely useful in the study of ultrastabilized zeolites, i.e. zeolites where aluminium has been removed from the framework. Since the removed aluminium remains in the pores of the zeolite, and no longer forms part of the lattice structure, the total sample composition is obtained from elemental analysis, whereas NMR yields the framework composition. Since the substitution of gallium for aluminium gives rise to similar (even larger) shifts of the ^{29}Si resonance[61-63], incorporation of gallium may be monitored in the same manner. The sensitivity of the technique to subtle structural features is illustrated by the observation of a splitting of the Si(1Ga) signal; this is probably due to differentiation between the (four possible) Si—O—Ga bridges that can be formed[61]. Many studies have been devoted to the ultrastabilization process of faujasite[64-69], with particular attention to the question of whether aluminium is located at specific sites in the unit cell and how aluminium is distributed at intermediate stages of dealumination[70-75]. From the intensity ratios of the silicon resonances it was concluded that, in addition to obeying the Loewenstein rule[1], some degree of ordering occurs, and several ordering schemes have been proposed. Mordenite, ZSM-5 and other zeolite dealumination processes have been monitored[70-81]. Typical intermediate spectra for mordenite ultrastabilization are shown in Figure 10, which also illustrates two additional aspects of the use of silicon NMR for zeolite characterization. The first is that multiple resonances occur after full dealumination. These arise from silicons in the four crystallographically different sites present in mordenite, and we shall discuss this more extensively below. Secondly, the inset shows that on cross-polarization a resonance at ca -100 ppm is specifically enhanced. From this enhancement and the resonance position, it is concluded that this resonance belongs to Si(OH) (Q^3) groups present at lattice defects formed during the ultrastabilization process. Such defects are found for most zeolites investigated during dealumination[65,81-84], and probably belong to those sites where aluminium has been removed from the framework and no silicon replacement has yet taken place. Quantification of these defects has been attempted, both by NMR and by other techniques, and appears to differ between preparations.

For ZSM-5, a large number (ca 20% of total silicon) of defects have been found in the as-synthesized sample when the structure-directing tetrapropylammonium ion is still present in the pores[85]. An alternative explanation for these defects has been suggested[86]. Extensive investigations on template-free ZSM-5[84-86] have indicated that the number of defects decreases with increasing aluminium content and that they may be repaired by a mild steaming treatment. Both dealumination and the restoration of these defects result in dramatic narrowing of lines in the ^{29}Si spectrum. Indeed, a number of studies devoted to the origin of linewidths have indicated that the major line-broadening factor is chemical shift dispersion due to small deviations from a perfectly crystalline structure or from perfect Si/Al ordering[87-90]. For 'perfect' (dealuminated) samples, the linewidths may be as narrow as 5 Hz[91]. These samples show a

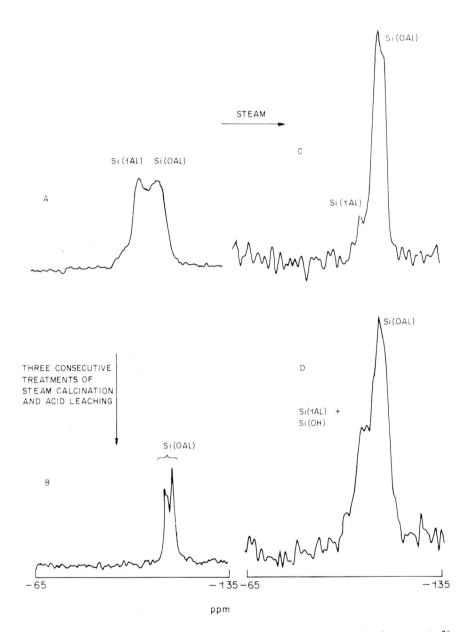

Fig. 10. ^{29}Si MAS NMR spectra monitoring the ultrastabilization of NH_4^+-mordenite[76]. (A) Starting sample; (B) sample after three consecutive steps of steam and acid treatment; (C) sample after one steam treatment; the Si (1 Al) signal is strongly reduced; (D) Sample C, spectrum recorded using CP

multitude of resolved ^{29}Si resonances, belonging to crystallographically distinct sites. Indeed, the dealumination procedure has been exploited to obtain ultra-high resolution, so as to permit structural characterization by NMR. Several different ways of relating ^{29}Si chemical shifts with the average geometry of Si—O—Si bonds have been proposed[92-95]. A compilation of these is shown in Figure 11. The full lines are least-squares lines derived from experimental data for well characterized silicates. The choice of the functions sec θ and $1/(1 - \cos \theta)$ is based on the theoretically predicted relationships of these functions with the hybridization index and fractional s character of the oxygens. As observed from Figure 11, in the regions where most Q^4 resonances occur, all proposed relations are more or less equivalent. The assignment of ^{29}Si resonances to specific crystallographic sites may be based on these relations. The accuracy and usefulness of assignment depends critically on the accuracy with which the Si—O—Si angle can be determined from XRD. It is clear that for complicated structures such as ZSM-5 (having 24 distinct crystallographic sites of equal occupancy) this will be a limiting factor. Since the average Si—Si distance is strongly correlated with the Si—O—Si angle and is more precisely determined from XRD-data, the Si—Si distance forms an alternative parameter to correlate chemical shift and geometry[96,97].

Benefiting from the high resolution obtained for high-silica zeolites, one could also demonstrate that zeolites are by no means inflexible structures. Dramatic sorbate-specific spectral changes were observed after absorption of benzene, acetylacetone, xylene or pyridine in ZSM-5[91,98,99]; similar effects were observed with temperature variation[100,101]. These changes reflect alterations in bond angles and distances in the structure. Recently zeolite rho framework modifications were also observed, depending on the state of hydration or interaction with methanol[102,103], divalent cations may have an equivalent effect[104].

(b) $^{27}Al\ NMR$

^{27}Al NMR has been used mainly in conjunction with ^{29}Si NMR for monitoring dealumination processes and for the detection of extra-lattice aluminium[68,69,105-108], since octahedral (extra-lattice) and tetrahedral aluminium resonate at ca 5 and 60 ppm respectively. Distinctions between several tetrahedral sites were made by Scholle and co-workers[109,110] for as-synthesized ZSM-5 and by Fyfe and co-workers[12,11] for strongly dealuminated ZSM-5 and zeolite omega. Particularly, for low Si/Al ratios, where the Si(Al) resonance is very weak in the ^{29}Si spectra, incorporation of aluminium in the framework can readily be demonstrated via ^{27}Al NMR. Extensive early investigations are reported in Refs 112 and 113. Until recently it was assumed that all extra-lattice aluminium would be octahedral. It has clearly been shown, however, that tetrahedral Al(OH)$_4$ is formed on calcination of Ca A, and this is easily recognized in the ^{27}Al spectrum by its chemical shift (ca 78 ppm)[114]. ^{27}Al NMR

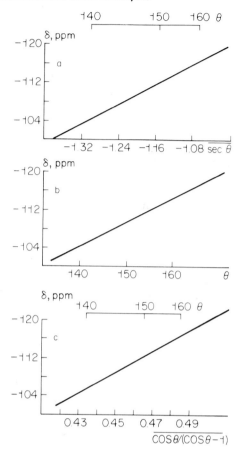

Fig. 11. Summary of suggested correlations between ^{29}Si chemical shifts and average Si—O—Si angle. (a) From Ref. 92; (b) from Ref. 94; (c) from Ref. 93

has also been used to demonstrate the effectiveness of realumination of ZSM-5[115,116]. ^{27}Al NMR spectroscopy had been hampered by large linewidths and field-dependent observed chemical shifts, even with careful experimentation at high fields. The distinction between Al species having the same chemical shifts and strongly overlapping lines but different quadrupolar coupling constants can be made via the nutation method. Using this method Geurts et al.[117] inferred that there should be three types of aluminium in ZSM-5. Zero-field NMR (see Section 2.2.5) may help to resolve lines in cases where high-resolution MAS experiments fail. If MAS can resolve lines, useful information can be derived. Lippmaa et al.[31] demonstrated that accurate chemical shifts can be determined by correcting the observed chemical shifts for the quadrupolar induced shift. To accomplish this, four different methods were employed, using either data from the ($\frac{1}{2}$ $-\frac{3}{2}$) transition, observed shifts at different magnetic fields, simulation of MAS lineshapes or nutation spectra. As a result, it could be concluded that the chemical shift of aluminium is, in a similar manner to silicon, sensitive to its environment: Q^3 aluminium resonances are found ca 10 ppm downfield from those of Q^4 aluminiums, and for Q^4 aluminium a correlation between chemical shift and average Al—O—Si angle (θ) was established:

$$\delta = -0.50\theta + 132 \text{ ppm} \tag{6}$$

with a correlation coefficient of 0.95.

Five coordinated aluminium resonates between the tetrahedral and octahedral positions at ca 35 ppm[31,118]. The quadrupolar coupling constant in aluminoscilicates varies between 1 and 6 MHz[31,119]. With this information available, the use of very-high-field ^{27}Al NMR will probably become increasingly important in the near future.

(c) ^{11}B NMR

As boron is one of the possible candidates for substitution in the tetrahedral sites in zeolites, it is obvious that ^{11}B NMR should be tested as a tool for structural characterization. Just as ^{27}Al and ^{17}O, ^{11}B is a quadrupolar nucleus. Particularly with boron the utilisation of VAS (see Section 2) has been demonstrated to yield good results[120]. The rationale for this is that the two types of boron coordination that are important in this context (tetrahedral BO_4 and trigonal BO_3) have similar isotropic chemical shifts but substantially different quadrupolar coupling constants. In addition, chemical shift anisotropy, which is essentially only averaged out by MAS, does not contribute to a large extent to the linewidth. Thus VAS allows the quadrupolar interaction to act as a shift parameter, shifting the resonance of the trigonal species away from its isotropic chemical shift position, thereby allowing discrimination between the two species.

Although boron NMR has been mostly applied to borosilicate glasses, a few examples of applications to zeolitic materials are available[121,122]. In an extensive study of boron-substituted silicalite (boralite) it was demonstrated that the highly symmetric tetrahedral boron severely distorts on dehydration, and probably changes to a three-fold coordination[122] (see Figure 12).

(d) ^{17}O NMR

Following ^{17}O enrichment, the oxygen nucleus is also available for study by NMR. A variety of zeolites and related materials have been studied, including gallozeolites and zeolitic aluminophosphates[123,124]. Differences in chemical shift and quadrupolar coupling constant have been found; for example, oxygens

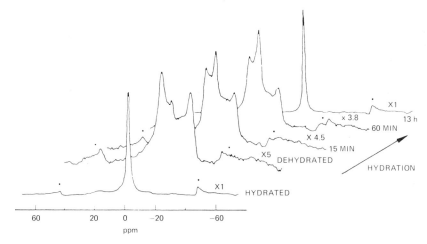

Fig. 12. ^{11}B spectra of boralite as a function of water content as 96.3 MHz. The asterisk denotes spinning sidebands of the NMR line of hydrated sites; the quadrupole pattern is due to dehydrated boron sites. From Ref. 122

bonded to two silicons and those bonded to one silicon and one aluminium have chemical shifts of ca 50 and ca 35 ppm with quadrupolar coupling constants of ca 5 and ca 3.2 MHz, respectively. Even when working at the highest available field, however, the quadrupolar interaction is too large to resolve the difference species in MAS or VAS spectra. They can be differentiated in static spectra on the basis of the large difference in quadrupolar constant, by simulation of the spectra. The experimentally observed magnitude of the quadrupolar coupling is in accordance with theoretical predictions; the lower experimental value for Si—O—Al and Si—O—Ga is rationalized as being due to cation coordination.

3.1.2   Characterization of other catalyst materials

(a) Clays

Naturally occurring clays (see Section 1 and Figure 5) are often complicated materials, owing to the large variability in the ionic composition of the octahedral layer and the interlayer[125–127]. The tetrahedral layer usually only contains Si, Al and O, with silicon-to-aluminium ratios varying between 1 and infinity. A second complicating factor is the occurrence of mixed-layer clays, consisting of stacked layers of different compositions. We shall discuss here studies on relatively well defined clays, which have provided the reference ^{29}Si NMR data collected in Figure 9[9,128–138], from which it is clear that for clays too, different environments of the tetrahedral SiO$_4$ can be discriminated in the NMR spectrum so that the degree of aluminium substitution in the tetrahedral layer can be determined in an analogous fashion to equation 5:

$$Si/Al_{tetr} = \sum_{n=0}^{n=3} I_{Si(nAl)} \Bigg/ \sum_{n=0}^{n=3} 0.33 n I_{Si(nAl)} \tag{7}$$

[27]Al spectra, in which the resonances of tetrahedral and octahedral Al are observed at ca 72 and ca 5 ppm, respectively, yield the ratio between these two differently coordinated ions; from this ratio and the general equation for the clay (equation 1), the degree of substitution in the tetrahedral layer can also be derived. It has been noted that the ratios calculated from [29]Si and [27]Al data are not always in accordance with each other[128,129]. This may be due to the fact that part of the aluminium is in a very asymmetric (five-coordinated?) environment, and is broadened beyond detection[129]. This situation should be avoided by using very high fields for [27]Al measurements and taking the appropriate experimental precautions (see Section 2.1.3). Another suggested explanation is the presence of some amorphous materials, which would cause a difference between the elemental composition of the clay and the overall sample composition[128,139]. Indeed, for high-field [27]Al data for a large set of well defined samples, the [29]Si- and [27]Al-derived degrees of substitution in the tetrahedral layers are reasonable consistent[125]. As for zeolites, the ordering in tetrahedral layers has been investigated. Since the experimentally observed intensity ratio of the Si(nAl) resonances differs from that expected for a statistical distribution of aluminium (taking into account the Loewenstein rule[1]), it was concluded that ordering takes place[129–131,140,141], and this appears to be correlated with the location of counter ions in the interlayer[141]. The [29]Si linewidths of clays with ordered Al distributions or without Al substitution in the tetrahedral layer are always smaller than those of unordered materials, in accordance with the observations for zeolites. The [29]Si resonances shift (within their specific ranges, see Figure 9) to low field with decreasing Si/Al_{tetr} ratio; the tetrahedral [27]Al resonances display the same behaviour[130]. For aluminium-free tetrahedral layers, multiple lines of the Si(OAl) resonance were observed[132,139,142]. These different resonances probably belong to sites having a different interaction with interlayer molecules or ions. Indications for this are found in the observation that the resonance of kaolinite shifts ca 2 ppm with different swelling solvents[133]; dehydration of 2:1 layer silicates was found to give rise to 2–10 ppm shifts[128,134]. ^{27}Al NMR has been of great importance in the study of the effectiveness and mechanism of pillaring of clays[136,139,143]. It could be demonstrated that on Al_{13} pillaring of beidellite clay, the pillar material reacts with the layer structure, whereby a link is probably formed between a tetrahedral aluminium of the clay layer and an octahedral aluminium of the pillar. The so-formed new high-surface-area material can be seen as a two-dimensional zeolite. No such reaction takes place with smectite clays that have no tetrahedral aluminium substitutions.

(b) Aluminophosphates

Since the first announcement of the synthesis of these porous crystalline materials[8] a number of NMR investigations for structural characterization have been performed. The presence of phosphorus induces a shift of the ^{27}Al resonance of about 20 ppm to high field[144]; this is found for both octahedral and tetrahedral aluminophosphate. The ^{27}Al shift and lineshape are strongly dependent on the specific crystalline structure studied. The observed chemical shift of the tetrahedral aluminium varies between 29 and 46 ppm[145–148]; the lineshapes sometimes show the patterns of unaveraged second-order quadrupolar interaction and some impurities of octahedral aluminium phosphate may be observed[145]. The ^{31}P signals are found between -24 and -31 ppm. A relationship between chemical shift and Al—O—P angle for both phosphorus and aluminium has been proposed[147]. In an extensive study by Blackwell and Patton[145] it was demonstrated that interaction with an organic template molecule or water strongly influenced the spectrum and the presence of such interactions was confirmed by the cross-polarizability of some peaks. The neutral aluminophosphate framework is of minor interest as a catalyst. However, a multitude of substitutions are possible (see Ref. 149 for an overview), again yielding a charged framework. A multinuclear NMR study of boron-substituted aluminophosphate proved that the boron was present in the framework, having replaced both aluminium and phosphorus[121].

(c) Alumina, silica and aluminosilicate

The surface structure of silica and the modifications thereof generated on dehydration and rehydration have been extensively investigated[150–154]. These investigations involved, among others, the probing of the surface accessibility using organic silane reagents as probes. Extensive studies of the organic materials that were immobilized on the surface have also been carried out, but we consider this subject to be outside the scope of this chapter. The silica surface can be characterized in terms of the number of geminal and single hydroxyl groups (corresponding to the Q^3 and Q^2 resonance in the ^{29}Si spectrum). Geminal hydroxyl groups appear to react more readily with silane reagents than isolated hydroxyls[150]. It was demonstrated that very high temperatures ($>1000\,^\circ$C) are needed to achieve full dehydroxylation of the silica[147]. Dehydration results in condensation of silanol groups, and this is observed as a decrease in Q^3 and Q^2 resonances and line broadening, indicative of lattice strain. Strained bonds generated at 500 °C (or lower) are the most easily hydrolysed on rehydration. Dehydration and rehydration is not a reversible process[153]. The NMR results indicate that the silica surface is heterogeneous, and contains segments resembling the 100 and 111 faces of cristabolite[153]. Relative quantification of the different surface species can be done, as in the former studies, using

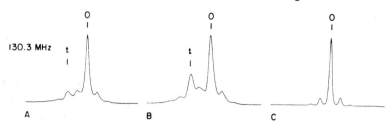

Fig. 13. 130.3 MHz ^{27}Al MAS solid-state NMR spectra of three alumina samples. (A) Bayerite after calcination at 270 °C; (B) bayerite after calcination at 470 °C; (C) boehmite after calcination at 1000 °C. From Ref. 155

CP ^{29}Si NMR. If data on the absolute number of silanol groups are required (as a percentage of total silicon), spectra need to be recorded without CP, using long recycle delays (ca 100 s). It was demonstrated that silica may contain 44% of silanol groups[154].

Several NMR investigations on the characterization of alumina and aluminosilicates have been reported. John et al.[155] determined the ratio of octahedral to tetrahedral aluminium for several transitional states of alumina (generated by treatments at different temperatures). They demonstrated that quantification can reliably be done from the ^{27}Al spectra obtained at high field in spite of severe peak overlap, the observed ratios being in accordance with X-ray structure predictions (see Figure 13).

McMillan et al.[156] confirmed the previous results and demonstrated that AlPO$_4$ and not Al$_2$(MoO$_4$)$_3$ is formed on a hydrodesulphurization catalyst made by impregnation of γ-alumina with NiMoP. Using ^{27}Al NMR, amorphous aluminosilicate samples with the same Si/Al ratio were shown to have different tetrahedral and octahedral aluminium occupancies, reflecting differences in the method of preparation. NMR spectroscopy is probably the only way to obtain this type of information[157]. Recently, a preliminary study of the dehydration behaviour of silica, alumina and aluminosilicate was performed using high-resolution proton NMR[158]. Brönsted acid sites associated with different surface species were observed, in addition to non-acidic OH groups (see also Section 4.2.2). For fluorinated alumina and aluminosilicate (which are important industrial catalysts) the formation of SiF and AlF bonds could be demonstrated from ^{19}F measurements[159].

3.2 Zeolite synthesis

Zeolites are sythesized by hydrothermal treatment of aqueous solutions of silica, containing the appropriate amount of added alumina, organic and inorganic bases, and adjusted to the correct pH. Many studies of such solutions have been undertaken in order to understand the mechanism of the synthesis. The first extensive NMR studies were reported by Hoebbel and co-workers[160,161] and Harris et al.[162]. The NMR spectra were found to contain broad lines belonging to a diversity of oligomers (the large polymeric species are unobserved in liquid-state NMR) which are centred in four separate spectral regions depending on the degree of polymerization of the silicon nucleus giving rise to the signal. These regions correspond to the Q^0, Q^1, Q^2 and Q^3 regions already mentioned in the discussion of solid-state spectra (Section 3.1.1). In 1971, Hoebbel and Wieker[163] reported the existence of well defined oligomeric silica species in solutions, stabilized by the addition of tetramethylammonium ions. The double four ring (D4R) oligomer is observed in ^{29}Si NMR spectra as a sharp signal in the Q^3 region[160]. Additional studies on oligomers were published, and these identified several specific oligomeric structures[164–167]. Groenen et al.[168] utilized the fact that the presence of such oligomers could be enhanced by the addition of organic solvents to the aqueous solution, to study the relative stability of D3R, D4R and D5R as a function of pH and temperature (see Figure 14). Assignment of resonances to distinct small oligomers was facilitated by the use of two-dimensional experiments of ^{29}Si enriched silicate solution[166,169].

The kinetics of interconversion of these species have been shown to be slow at room temperature using time-dependent studies and saturation transfer ^{29}Si NMR[170,171]. At the higher temperatures the equilibrium shifts to a lower degree of polymerization, although a small amount of double ring oligomer is still present at 100 °C[168]. Recently, an extensive study of oligomeric species in solutions was reported in which many resonances in the liquid state spectrum were assigned to additional individual oligomeric species. This was made possible by the use of very high magnetic field NMR instrumentation (500 MHz)[172].

Naturally the incorporation of aluminium in these oligomers has been a major point of interest. In 1982 Engelhardt et al.[161] identified the D4Rs with varying aluminium content from the ^{29}Si NMR data and gas chromatographic (GC) analysis of silylated species. They concluded that aluminous double four rings exist as a stable species in solution. From these solutions D4R crystals, some with very high aluminium contents, were derived. The incorporation of aluminium in the D4Rs was demonstrated via solid-state ^{29}Si NMR. Incorporation of other ions in double ring oligomers in solution was also demonstrated[169,173].

Advancement in the mechanics of magic angle spinning offers a very attractive opportunity in this field. The possibility of studying the synthesis solutions (gels) under synthesis conditions (increased temperature and pressure) would provide a clearer understanding of the synthesis mechanism. MAS studies of aluminosilicate gels have been reported[174–177]. Intermediate stages of ZSM-5[178] and zeolite A synthesis were investigated using MAS[176,177]; in ZSM-5 synthesis, an aluminium-rich phase and small ZSM-5-type complexes with the template were the first solid materials found, and for zeolite A synthesis it was concluded that

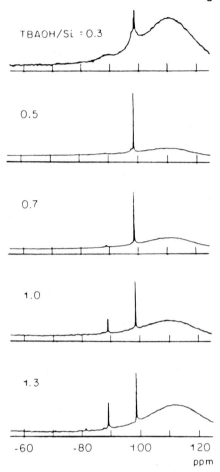

Fig. 14. ^{29}Si NMR spectra of solutions prepared from tetramethoxysilane in TBAOH–H$_2$O after addition of DMSO (50 % v/v) as a function of the TBAOH/Si ratio; the effect of pH is clearly observed

an amorphous gel consisting of tetrahedral Si(OAl)$_4$ and Al(OSi)$_4$ building units was formed initially. The type of species formed appears to be highly dependent on the method of preparation, and probably much work still needs to be done in order to obtain a deeper insight into the mechanism of zeolite synthesis.

3.3 Cations in zeolites

The excess negative charge of the framework, due to the presence of Al or other substituent metal ions, is balanced either by protons (which form the Brönsted acid sites) or by other exchangeable cations. These ions are located as well defined sites in the zeolite, but may also be very mobile. Just as the acid sites (see Section 4), these can be studied via ^{15}N NMR using NH_3 or acetonitrile as a probe. The observed ^{15}N shift of ammonia is proportional to the number of cations[179,180], while the acetonitrile resonance is particularly sensitive to the electrostatic potential of the cation in question[181]. An indirect but very elegant demonstration of the very high cation mobility in hydrated zeolite A was obtained from ^{29}Si NMR[182]. The silicon chemical shifts of Na A and Li A differ by about 4 ppm, but mixtures of the two materials give rise to one signal at the averaged resonance position, indicative of rapid migration and mixing of the cations.

Cations in zeolites that have been the subject of NMR studies are ^7Li, ^{23}Na, ^{105}Tl and ^{133}Cs. Apart from thallium, these are all quadrupolar nuclei with the associated difficulties in measurement. The principal questions that have been addressed in these studies are the mobility of the cations as a function of temperature and degree of hydration, and the location of the ions as monitored by the components of the electric field gradient tensor[183,184]. Relaxation measurements have sometimes been hampered by the presence of small amounts of paramagnetic impurities[185,186]. The studies on thallium[187,188] in Na A revealed at least two Tl sites and the mobility of the Tl ions appears to depend on both the water content and the nature of Tl ions at the other cation sites (Na or Tl). It has been shown that it is also possible for quadrupolar nuclei to discriminate between different sites: using MAS, two different Cs sites were observed in Cs-exchanged mordenite and these could be assigned to sites known from crystallographic studies[189]. Similarly, assignments could be made for Na in zeolite A on the basis of nutation NMR spectra[190].

3.4 Probing the pore geometry of zeolites using Xe NMR

One of the main incentives in the investigation of zeolite structures is the elucidation of zeolite-specific selectivity in catalysis. This arises because of differential diffusion of molecules with different sizes and shapes in the zeolite pores. It would therefore appear extremely instructive to probe the pore structure directly, with a molecule that 'observes' the zeolite surface as would a reactant. ^{129}Xe was shown to be a very suitable and sensitive nucleus for this purpose. The extended Xe electron cloud is easily deformable (e.g. due to collisions), and deformation results in a large low-field shift of the Xe resonance. Hence, assuming fast exchange, the shift of Xe absorbed in zeolites can be regarded as the sum of several additive contributions[191]:

$$\delta = \delta_0 + \delta_{Xe} + \delta_E + \delta_S \qquad (8)$$

where δ_0 is a reference (Xe gas at infinitely low pressure), δ_{Xe} is a contribution from Xe–Xe collisions, δ_E results from electric field gradients (cations) and δ_S is due to collisions with the zeolite. From this equation it is clear that, when extrapolated to vanishing Xe pressure, the Xe chemical shift is determined solely by the electrostatic term (which can be shown to be negligible for monovalent ions[191,192]) and the Xe zeolite interaction. Thus, Xe NMR has been used in the study of zeolites X, Y, L, A, ZSM-5, ZSM-11 and mordenite, both in the H-state and with monovalent and divalent cations[191–196]. Indeed, the Xe chemical shift increases with decreasing mean free path for the Xe atoms in the pores[195,196]. As an alternative explanation for the sensitivity of the Xe shift to the pore diameter, it was suggested that δ could be determined by the surface curvature of the pore walls[197]. If more than one adsorption site is present in zeolites, and exchange between sites is slow, multiple Xe resonances may be observed. This was shown to be the case for Na-mordenite[193,195], where Xe is trapped in the side pockets of the structure and the presence of Na^+ considerably slows down the exchange rate (see Figure 15). The resonance shift of the side-pocket signal is independent of Xe pressure and this indicates that only one Xe atom interacts with the side-pocket surface, and no Xe–Xe interactions play a role[193,195].

Divalent cations in zeolites do have an effect on Xe shifts. Indeed, interaction of these cations with Xe is very strong, so that Xe adsorbs preferentially at these cations[198,199]. This gives rise to a non-linear dependence of the Xe chemical shift on the Xe loading, allowing the determination of δ_E. A relationship between δ_E and the electronegativity of the cations has been proposed[199]. The method opens up a way to determine the location of cations, i.e. whether or not they are accessible for Xe. Further, [129]Xe NMR can be used in the study of metallic particles in zeolites[200], while reduction–oxidation reactions can also be monitored[201].

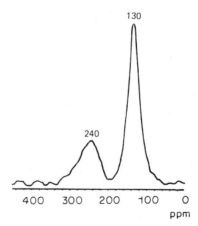

Fig. 15. [129]Xe spectrum of xenon adsorbed on Na-Mordenite

4 CATALYST ACIDITY

The acidic properties of a catalyst frequently play a crucial role in the catalytic action and it is for this reason that many investigations have specifically examined intrinsic acidity properties. For zeolites, as we have mentioned in Section 3.1, ^{29}Si NMR has played an essential role in the determination of the amount of aluminium (or other ions such as boron or gallium) that actually forms a part of the zeolite framework. The distribution of aluminium over distinct crystallographic sites was also derived in this manner (see Section 3.1).

For highly siliceous zeolites, ^{27}Al NMR proved to be an additional tool. The distinction between framework (tetrahedral) and non-framework (mainly octrahedral) aluminium is easily made from sufficiently high-field ^{27}Al NMR spectra. While ^{29}Si and ^{27}Al NMR are suitable for determining the concentration of acid sites, the actual acid strength can be probed either by direct study of the acid protons (Brönsted acid sites) or by using a probe molecule. These two methods will be discussed in the next sections.

4.1 Direct observation of Brönsted acid sites

Solid-state ^{1}H NMR is an experimentally difficult technique, particularly in situations where there is a high concentration of protons, and the proton–proton interactions which cause extreme line broadening have to be removed by multiple quantum decoupling. Fortunately for dehydrated zeolites, the proton density is often so low that utilization of this technique is not always required. However, residual broadening still severely limits the information content of proton spectra because of the small chemical shift dispersion of ^{1}H (compared with, say, ^{13}C). For aluminosilicates, Pfeifer and his group have been the pioneers of the use of ^{1}H NMR. They succeeded in overcoming the difficulties of MAS of samples sealed in glass capillaries[202] and established reference values for the chemical shifts of hydroxyl groups in zeolites and other carrier materials with or without adsorbed pyridine. A review of NMR studies on zeolite acidity is given in Ref. 203. Extensive studies have been carried out for zeolite H-Y, where four different ^{1}H resonances are found and assigned to residual NH_4, two different acid sites and non-acidic SiOH groups[204,205]. The two different acid sites were correlated with two OH frequencies observed in the IR spectrum, and assigned to acidic OHs in the large and small cavities in the framework. The thermal dehydroxylation process could be characterized as being necessarily accompanied by dealumination[204]. Further, a correlation could be establish between acid strength and the chemical shift of the resonance of the acidic OHs in the large cavity[205]. Studies of ZSM-5 revealed that here too Brönsted acid sites may be differentiated from non-acidic SiOH[205,206]. The acidity of the boron analogue of ZSM-5, boralite, was demonstrated to be less than that of ZSM-5[207].

4.2 The use of probe molecules to study acid sites

4.2.1 Zeolites

Acid sites in zeolites can be sensitively probed by the study of adsorbed molecules such as pyridine or ammonia using IR and NMR. In the early NMR work, the monitoring of pyridine adsorption was performed via ^{13}C NMR; however, it was apparent that the ^{15}N resonance of pyridine and ammonia is much more sensitive to differences in environments of adsorption. For ^{15}N studies, close to 100%-enriched adsorbents need to be used to improve the poor signal-to-noise ratio, particularly at low loading levels. Valuable data have been obtained for zeolite H-Y[179,180] where, at high loading levels, all added ammonia appears to be converted to ammonium ions owing to interaction with Brönsted acidic hydroxyl groups. A substantial difference in ^{15}N chemical shift is found for ammonia interacting with Lewis or Brönsted acid sites. Interaction of ammonia with Na ions gives different chemical shifts (see Section 3.3), allowing the determination of the number of these sites. The above-mentioned work relies on fast exchange between molecules adsorbed at acid sites and those 'physisorbed' in the pores. At low loading levels, where all molecules are adsorbed on the acid sites, the signals are broadened beyond detection as a result of immobilization of the absorbed molecules. In order to circumvent the problem, CP-MAS techniques are needed. Ripmeester[208] was able to observe two different resonances of immobilized pyridine molecules in acid-leached, calcined mordenite and these were ascribed to Lewis and Brönsted acid sites (in addition to a resonance assigned to physisorbed pyridine). A new approach was followed by Rothwell et al.[209], who explored the utilization of the much more sensitive ^{31}P nucleus as a probe for acidity. Trimethylphosphine indeed proved an extremely suitable probe molecule. Direct proof of protonation of the phosphorus on Brönsted acid sites was found in the observation of ^{1}H–^{31}P J coupling in the MAS spectra of adsorption complexes with H-Y zeolite (Figure 16). Depending on the temperature of calcination, a number of different resonances of Lewis sites were found and one of these could be assigned to absorption on extra-lattice Al_2O_3[210].

4.2.2 Alumina

Surface acid sites on alumina have been studied using techniques similar to those used for zeolites. The CP-MAS ^{15}N data of Ripmeester[208] (pyridine adsorbed on γ-alumina) reveal the presence of two different Lewis acid sites (pyridine is not basic enough to adsorb at Brönsted acid sites), one of which is

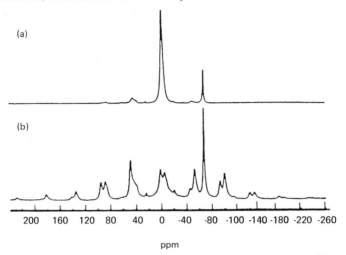

Fig. 16. (a) High-power, proton-decoupled ^{31}P MAS NMR spectrum of trimethylphosphine adsorbed on H-Y zeolite. (b) Proton coupled spectrum. The ^{1}H–^{31}P J coupling is evident. From Refs 209, 217 and 220

sensitive to the presence of water. Exchange of these immobilized pyridine molecules with the more mobile species was very slow. From an extensive ^{15}N study, complemented with ^{2}H data[211], it was found that the two observed Lewis acid sites can be assigned to octahedral and tetrahedral Al. ^{2}H investigations indicated that there is a distribution of mobilities of the adsorbed pyridine and that these data may be used to probe the steric properties of the binding sites. In contrast to pyridine, ammonia adsorbs at both Lewis and Brönsted acid sites. Although the ^{15}N signal is sensitive to the difference in sites, the resolution is not sufficient to make discrimination between sites possible. Such a discrimination can be made, however, from ^{2}H NMR data of adsorbed N^2H_3. The use of trialkylphosphines as probes to study the acidity of alumina[210,212,213] or silica–alumina[213] carriers led to the detection of Lewis acid sites on alumina and both Lewis and Brönsted types of sites in silica–alumina.

5   ADSORBED MOLECULES

5.1   Molecular mobility

Several methods have been applied to study the mobility of substrates in catalysts. The most obvious parameters which give an insight into molecular mobility are the relaxation times (T_1, $T_{1\rho}$, T_2). An excellent overview of the use of relaxation measurements for studying adsorbed species was given by Pfeifer[214].

Another related parameter is the lineshape of the resonance of a quadrupolar nucleus (particularly ^{2}H), which may also yield the anisotropy of molecular motion. One of the new advances in this area is the two-dimensional approach, both in ^{13}C NMR (see Ref. 215 for a theoretical treatment of this technique) and ^{2}H NMR (see, e.g., Ref. 216) from which direct information of modes of motion can be derived. Faster translational motions may be studied using (pulsed) field gradient techniques in which the actual displacement of a molecule is measured.

For all types of measurement of motion mentioned in this section, interpretation usually relies on comparison of experimental results with detailed theoretical calculations and simulations. We foresee that such calculations will become increasingly important in this field of research.

Rather than covering the whole of the literature, the following examples were chosen from recent publications to illustrate the present state of the art, and the scope of these distinct techniques (for a review of the NMR of organic molecules adsorbed to porous solids, see Ref. 11, 14 and 217; Ref. 11 also gives a good overview of relaxation measurements on sorbed water).

For studying the dynamics of small molecules in the pores of zeolites, a variety of approaches have been chosen, and generally yield complementary information. From pulsed field gradient measurements, it was concluded that the translational mobility of benzene, toluene and xylene in zeolite X is limited by jumps through the windows of the zeolite lattice[218]. The mobility of such molecules in zeolites was shown to be influenced by co-adsorption and changes of the catalyst during operation (e.g. coke deposition)[219].

The rotational dynamics of aromatic molecules have been probed by ^{2}H and ^{13}C measurements. Using ^{2}H NMR, Eckman and Vega[220,221] were able to observe the onset of rotation of p-xylene in ZSM-5 around its long axis with increasing temperature, and concluded that reorientation of the long axis does not occur. Therefore, p-xylene would not diffuse in the zig-zag channels of ZSM-5. In contrast, from ^{13}C results, isotropic diffusion of p-xylene in ZSM-5 was proposed[222]. A combination of both measurements under identical experimental conditions (e.g. loading) would be needed to reconcile both sets of results. ^{13}C measurements also indicated that the C6 axis of benzene, adsorbed to silicalite, undergoes rapid jumps between distinct orientations[223], in accordance with the conclusions from ^{2}H measurements, which, however, indicated such a motion for only a fraction of the loaded benzene[221]. Similar experiments were performed to investigate the influence of loading level on the state of adsorption of dimethyl ether on zeolite H-rho[224]. A diversity of motionally distinct adsorbed states was found. Using ^{13}C NMR, several different adsorption sites for methanol on the same zeolite could be identified[225]. A multinuclear study demonstrated distinct effects on framework structure caused by methanol or water adsorption[226] [see also Section 3.1.1(a)]. Water adsorption and competitive adsorption between water and polar or non-polar organic molecules can be conveniently studied via ^{1}H MAS NMR[227]. The determination of differential ordering characteristics of adsorbed molecules on alumina surfaces[228,229] and the properties of supported organometallic catalysts[230] are additional analytical applications of the NMR technique.

5.2   Active metal sites and reaction intermediates

The difficult work of investigating Pt catalysts has been addressed by Slichter's group (for review, see Refs 231 and 232). Platinum metal is the only metal used in catalysis that is reasonably accessible for NMR mesurements (spin-$\frac{1}{2}$, high γ and natural abundance). ^{195}Pt lineshapes (which under the experimental conditions are approximately 16 MHz in width) vary with Pt dispersion on the alumina carrier surface owing to reduced Knight shifts for Pt atoms located at the surface of a Pt particle. Using double resonance techniques (SEDOR: *S*pin *E*cho *DO*uble *R*esonance, where the surface Pt atoms can be specifically monitored when CO is adsorbed), it was demonstrated that CO binds to the metal via the carbon[233]. Using ^{13}C NMR the structure of adsorbed acetylene was concluded to be $\frac{3}{4}$CCH$_2$ and $\frac{1}{4}$HCCH. In addition, the structure and reaction intermediates of adsorbed ethylene could be unambiguously established[234]: ethylene is adsorbed as an ethylidyne group, while after chain scision mostly a naked carbon atom remains attached to the surface. ^{13}C studies of these molecules adsorbed to Rh, Os, Ir, Ru and Pd have been carried out and yielded similar results.[232]. Also in this area of research the availability of (newly designed) MAS equipment suitable for *in situ* studies offer many advantages[235]. The adsorption of CO on catalysts has been studied by a variety of groups. Duncan *et al.*[236] performed pioneering work on the discrimination between CO adsorbed on Rh clusters and isolated sites. MAS ^{13}C NMR can easily discriminate between linear, bridged and dicarbonyl adsorbed states of CO on Ru-Y[237]. Recently it was demonstrated that adsorption isotherms can be reliably obtained from NMR measurements, facilitating the interpretation of NMR results which depend on loading level[238]. The results of catalyst action can also be visualized: with carbonaceous deposits on a silica-supported Ru catalyst (a carbide on Ru and two distinct C-phases on silicon) the alkyl intermediates could be characterized using MAS[239,240].

6   FUTURE OUTLOOK

As is clear from the foregoing text, a wealth of information on zeolites, silicates and solid catalysts can be obtained from NMR investigations. This area of research has benefited enormously from recent progress in NMR techniques, which can now be harnessed to their full extent. The more general availability of high fields, higher spinning speeds, extended *in situ* capability and, not least, the progress in theoretical understanding and simulation methods indicates a promising outlook for the future. The opportunities to study quadrupolar nuclei at high or zero magnetic field will widen the scope still further in the ever expanding field of heterogeneous catalysis.

REFERENCES

1. W. Loewenstein, *Am. Mineral.*, **39**, 92 (1954).
2. F. Jankowski and K. Bergk, *Z. Chem.*, **22**, 277 (1982).
3. D. W. Breck, *Zeolite Molecular Sieves: Structure, Chemistry and Use*, Wiley, London, 1974.
4. R. M. Barrer, *Zeolites and Clay Minerals as Sorbents and Molecular Sieves*, Academic Press, London, 1978.
5. R. M. Barrer, *Hydrothermal Chemistry of Zeolites*, Academic Press, London, 1982.
6. W. M. Meier and D. H. Olson, *Atlas of Zeolite Structure, Types* 2nd revised edition Butterworth, London (1987).
7. R. Gramlich-Meier and W. M. Meier, *J. Solid State Chem.*, **44**, 41 (1982).
8. S. T. Wilson, B. M. Lok, C. A. Messina, T. R. Cannan and E. M. Flanigen, *J.Am. Chem. Soc.*, **104**, 1146 (1982).
9. E. Lippmaa, M. Mägi, A. Samoson, G. Engelhardt and A. R. Grimmer, *J. Am. Chem. Soc.*, **102**, 4889 (1980).
10. E. Lippmaa, M. Mägi, A. Samoson, M. Tarmak and G. Engelhardt *J. Am. Chem. Soc.*, **103**, 4992 (1981).
11. J. Klinowski, *Prog. Nucl. Magn. Reson. Spectroc.*, **16**, 237 (1984).
12. G. T. Kokotailo, C. A. Fyfe, G. J. Kennedy, G. C. Gobbi, H. Strobl, C. T. Pasztor, G. E. Barlow and S. Bradley, *Pure Appl. Chem.*, **58**, 1367 (1986).
13. C. A. Fyfe, J. M. Thomas, J. Klinowski and G. C. Gobbi, *Angew. Chem.*, **95**, 257 (1983).
14. J. B. Nagy, G. Engelhardt and D. Michel, *Adv. Colloid Interface Sci.*, **23**, 67 (1985).
15. J. M. Thomas and J. Klinowski, *Adv. Catal.*, **33**, 199 (1985).
16. R. L. Austermann, D. R. Denley, D. W. Hart, P. B. Himelfarb, R. M. Irwin, M. Narayana, R. Szentirmay, S. C. Tang and R. C. Yeates, *Anal. Chem.*, **59**, 68R (1987).
17. T. M. Duncan and C. Dybowski, *Surf. Sci. Rep.*, **1**, 157 (1981).
18. W. Derbyshire. In G. A. Webb (ed.), *Nuclear Magnetic Resonance*, Vol. 11, Specialist Periodical Report, Royal Society of Chemistry, London, 1982, p. 264.
19. A. D. H. Clague. In G. C. Bond and G. Webb (eds), *Catalysis*, Vol. 7, Specialist Periodical Report, Royal Society of Chemistry, London, 1985, p. 75.
20. M. Mehring, *Principles of High Resolution NMR in Solids*, Springer, New York, 1983.
21. C. P. Slichter, *Principles of Magnetic Resonance*, 2nd ed., Springer, New York, 1987.
22. U. Haeberlen, *High Resolution NMR in Solids: Selective Averaging*, Advances in Magnetic Resonance, Suppl. 1, Academic Press, New York, 1976.
23. A. Abragam, *The Principles of Nuclear Magnetism*, Oxford University Press, London, 1961.
24. A. Samoson, E. Kundla and E. Lippmaa, *J. Magn. Reson.*, **49**, 350 (1982).

25. D. Freude, J. Haase, J. Klinowski, T. A. Carpenter and G. Roniker, *Chem. Phys. Lett.*, **119**, 365 (1985).
26. D. Freude and H. J. Behrens, *Cryst. Res. Technol.*, **16**, K36 (1981).
27. H. J. Behrens and B. Schnabel, *Physica*, **114B**, 185 (1982).
28. D. Muller, *Ann. Phys.*, **39**, 451 (1982).
29. E. Kundla, A. Samoson and E. Lippmaa, *Chem. Phys. Lett.*, **83**, 229 (1981).
30. A. Samoson, *Chem. Phys. Lett.*, **119**, 29 (1985).
31. E. Lippmaa, A. Samoson and M. Mägi, *J. Am. Chem. Soc.*, **108**, 1730 (1986).
32. A. Samoson and E. Lippmaa, *Chem. Phys. Lett.*, **100**, 205 (1983).
33. S. Ganapathy, S. Schramm and E. Oldfield, *J. Chem. Phys.*, **77**, 4360 (1982).
34. D. Fenzke, D. Freude, T. Fröhlich and J. Haase, *Chem. Phys. Lett.*, **111**, 171 (1984).
35. A. Samoson and E. Lippmaa, *Phys. Rev. B.*, **28**, 6567 (1983).
36. A. P. M. Kentgens, J. J. M. Lemmens, F. M. M. Geurts and W. S. Veeman, *J. Magn. Reson.*, **71**, 62 (1987).
37. E. R. Andrew, A. Bradbury and R. G. Eades, *Nature (London)*, **18**, 1659 (1958).
38. E. R. Andrew, *Arch. Sci.*, **12**, 103 (1959).
39. I. H. Lowe, *Phys. Rev. Lett.*, **2**, 285 (1959).
40. E. R. Andrew, *Prog. Nucl. Magn. Reson. Spectrosc.*, **8**, 1 (1972).
41. W. T. Dixon, *J. Magn. Reson.*, **44**, 220 (1981).
42. M. A. Hemminga, P. A. de Jager, K. P. Datema and J. Breg, *J. Magn. Reson.*, **50**, 508 (1982).
43. M. A. Hemminga and P. A. de Jager, *J. Magn. Reson.*, **51**, 339 (1983).
44. E. Oldfield, S. Schramm, M. D. Meadows, K. A. Smith, R. A. Kinsey and J. Ackerman, *J. Am. Chem. Soc.*, **104**, 919 (1982).
45. S. Schramm and E. Oldfield, *J. Chem. Soc., Chem. Commun.*, 980 (1982).
46. J. S. Waugh, L. Huber and U. Haeberlen, *Phys. Rev. Lett.*, **20**, 180 (1968).
47. P. Mansfield, *J. Phys. E.*, **4**, 1444 (1971).
48. W. K. Rhim, D. D. Elleman and R. W. Vaughan, *J. Chem. Phys.*, **59**, 3740 (1973).
49. W. K. Rhim, D. D. Elleman, L. B. Schreiber and R. W. Vaughan, *J. Chem. Phys.*, **60**, 1595 (1974).
50. D. P. Burum and W. K. Rhim, *J. Chem. Phys.*, **71**, 944 (1979).
51. A. M. Thayer and A. Pines, *Acc. Chem. Res.*, **20**, 47 (1987).
52. D. B. Zax, A. Bielecki, K. W. Zilm, A. Pines and D. P. Weitekamp, *J. Chem. Phys.*, **83**, 4877 (1985).
53. P. R. Barron, R. L. Frost and J. O. Skjemstad, *J. Chem. Soc., Chem. Commun.*, 581 (1983).
54. D. J. Cookson and B. E. Smith, *J. Magn. Reson.*, **63**, 217 (1985).
55. M. Goldman, *Spin Temperature and Nuclear Magnetic Resonance in Solids*, Oxford Univeristy Press, London, 1970.
56. H. W. Spiess. In *NMR: Basic Principles and Progress* P. Diehl, E. Fluck and R. Kosfeld (Eds.) Vol. 15, Springer, New York, 1978, p. 55.
57. S. R. Hartmann and E. L. Hahn, *Phys. Rev.*, **128**, 2042 (1962).
58. C. A. Fyfe, G. C. Gobbi, J. S. Hartman, R. E. Lenkinski, J. H. O'Brien, E. R. Beange and M. A. R. Smith, *J. Magn, Reson.*, **47**, 168 (1982).
59. F. D. Doty and P. D. Ellis, *Rev. Sci. Instrum.*, **52**, 1868 (1981).
60. G. R. Hays, *Analyst (London)*, **107**, 1272 (1982).
61. X. Liu, J. Klinowski and J. M. Thomas, *Chem. Phys. Lett.*, **127**, 563 (1986).
62. J. M. Thomas and X. S. Liu, *J. Phys. Chem.*, **90**, 4843 (1986).
63. D. E. W. Vaughan, M. T. Melchior and A. J. Jacobson, *ACS Symp. Ser.*, No. 218, 231 (1983).
64. I. E. Maxwell, W. A. van Erp, G. R. Hays, T. Couperus, R. Huis and A. D. H. Clague, *J. Chem. Soc., Chem. Commun.*, 523 (1982).

65. G. Engelhardt, U. Lohse, A. Samoson, M. Mägi, M. Tarmak and E. Lippmaa, *Zeolites*, **2**, 59 (1982).
66. J. Klinowski, J. M. Thomas, C. A. Fyfe and G. C. Gobbi, *Nature (London)*, **296**, 533 (1982).
67. G. Engelhardt, U. Lohse, V. Patzelova, M. Mägi and E. Lippmaa, *Zeolites*, **3**, 233 (1983).
68. J. Klinowksi, C. A. Fyfe and G. C. Gobbi, *J. Chem. Soc., Faraday Trans. 1*, **81**, 3003 (1985).
69. P. J. Grobet, P. A. Jacobs, and H. K. Beyer, *Zeolites*, **6**, 47 (1986).
70. G. Engelhardt, E. Lippmaa and M. Mägi, *J. Chem. Soc., Chem. Commun.*, 712 (1981).
71. J. Klinowski, S. Ramdas, J. M. Thomas, C. A. Fyfe and S. Hartman, *J. Chem. Soc., Faraday Trans. 2*, **78**, 1025 (1982).
72. M. T. Melchior, D. E. W. Vaughan and A. J. Jacobson, *J. Am. Chem. Soc.*, **104**, 4859 (1982).
73. G. Engelhardt, U. Lohse, V. Patzelova, M. Mägi and E. Lippmaa, *Zeolites*, **3**, 239 (1983).
74. G. Engelhardt, U. Lohse, E. Lippmaa, M. Tarmak and M. Mägi, *Z. Anorg. Allg. Chem.*, **482**, 49 (1981).
75. S. Ramdas, J. M. Thomas, J. Klinowksi, C. A. Fyfe and S. Hartman, *Nature (London)*, **292**, 228 (1981).
76. G. R. Hays, W. A. van Erp, N. C. M. Alma, R. Huis and A. E. Wilson, *Zeolites*, **4**, 377 (1984).
77. P. Bodart, J. B. Nagy, G. Debras, Z. Gabelica and P. A. Jacobs, *J. Phys. Chem.*, **90**, 5183 (1986).
78. C. A. Fyfe, G. C. Gobbi and A. Putnis, *J. Am. Chem. Soc.*, **108**, 3218 (1986).
79. J. B. Nagy, Z. Gabelica, G. Debras, P. Bodart and E. G. Derouane, *J. Mol. Catal.*, **20**, 327 (1983).
80. C. A. Fyfe, G. C. Gobbi, J. Klinowski, A. Putnis and J. M. Thomas, *J. Chem. Soc., Chem. Commun.*, **556** (1983).
81. C. A. Fyfe, G. C. Gobbi and G. J. Kennedy, *Chem. Lett.*, 1551 (1983).
82. A. W. Chester, Y. F. Chu, R. M. Dessau, G. T. Kerr, and C. T. Kresge, *J. Chem. Soc., Chem. Commun.*, 289 (1985).
83. G. L. Woolery, L. B. Allemany, R. M. Dessau and A. W. Chester, *Zeolites*, **6**, 14 (1986).
84. R. M. Dessau, K. D. Schmidt, G. T. Kerr, G. L. Woolery and L. B. Alemany, *J. Catal.*, **104**, 484 (1987).
85. G. Boxhoorn, A. G. T. G. Kortbeek, G. R. Hays and N. C. M. Alma, *Zeolites*, **4**, 15 (1984).
86. J. Keijsper and J. Dorrepaal, N. C. M. Alma, in preparation.
87. C. A. Fyfe, G. C. Gobbi, W. J. Murphy, R. S. Ozubko and D. A. Slack, *J. Am. Chem. Soc.*, **106**, 4435 (1984).
88. C. A. Fyfe, G. C. Gobbi, G. J. Kennedy, C. T. de Schutter, W. J. Murphy, R. S. Ozubko and D. A. Slack, *Chem Lett.*, 163 (1984).
89. C. A. Fyfe, G. C. Gobbi, W. J. Murphy, R. S. Ozubko and D. A. Slack, *Chem. Lett.*, 1547 (1983).
90. J. M. Newsam, *J. Phys. Chem.*, **89**, 2002 (1985).
91. C. A. Fyfe, J. H. O'Brien and H. Strobl, *Nature London*, **326**, 281 (1987).
92. J. V. Smith and C. S. Blackwell, *Nature (London)*, **303**, 223, (1983).
93. G. Engelhardt and R. Radeglia, *Chem. Phys. Lett.*, **108**, 271 (1984).
94. J. M. Thomas, J. Klinowski, S. Ramdas, B. K. Hunter and D. T. B. Tennakoon, *Chem. Phys. Lett.*, **102**, 158 (1983).

95. R. Radeglia and G. Englehardt, *Chem. Phys. Lett.*, **114**, 28 (1985).
96. E. J. J. Groenen, N. C. M. Alma, J. Dorrepaal, G. R. Hays and A. G. T. G. Kortbeek, *Zeolites*, **5**, 361 (1985).
97. S. Ramdas and J. Klinowski, *Nature (London)*, **308**, 512 (1984).
98. G. A. Fyfe, G. J. Kennedy, C. T. de Schutter and G. T. Kokotailo, *J. Chem. Soc., Chem. Commun.*, 541 (1984).
99. G. W. West, *Aust. J. Chem.*, **37**, 455 (1984).
100. D. G. Hay, H. Jaeger and G. W. West, *J. Phys. Chem.*, **89**, 1070 (1985).
101. C. A. Fyfe, G. J. Kennedy, G. T. Kokatailo, J. R. Lyerla and W. W. Fleming, *J. Chem. Soc., Chem Commun.*, 740 (1985).
102. A. J. Vega and Z. Luz, *J. Phys. Chem.*, **91**, 365 (1987).
103. Z. Luz and A. J. Vega, *J. Phys. Chem.*, **91**, 374 (1987).
104. P. J. Grobet, W. J. Mortier and K. Van Genechten, *Chem. Phys. Lett.*, **119**, 361 (1985).
105. D. Freude, T. Fröhlich, H. Pfeifer and G. Scheler, *Zeolites*, **3**, 171 (1983).
106. J. Klinowski, J. M. Thomas, C. A. Fyfe, G. C. Gobbi and J. S. Hartman, *Inorg. Chem.*, **22**, 63 (1983),
107. V. Bosacek and V. M. Mastikhin, *J. Phys. Chem.*, **91**, 260 (1987).
108. D. Freude, T. Fröhlich, M. Hunger, H. Pfeifer and G. Scheler, *Chem. Phys. Lett.*, **98**, 263 (1983).
109. K. F. M. G. J. Scholle, W. S. Veeman, P. Frenken and G. P. M. van der Velden, *J. Phys. Chem.*, **88**, 3395 (1984).
110. K. F. M. G. J. Scholle and W. S. Veeman, *J. Phys. Chem.*, **89**, 1850 (1985).
111. C. A. Fyfe, G. C. Gobbi, J. Klinowski, J. M. Thomas and S. Ramdas, *Nature (London)*, **296**, 530 (1982).
112. C. A. Fyfe, G. C. Gobbi, J. S. Hartman, J. Klinowski and J. Thomas, *J. Phys. Chem.*, **86**, 1247 (1982).
113. J. B. Nagy, Z. Gabelica, G. Debras, E. G. Derouane, J. P. Gilson and P. A. Jacobs, *Zeolites*, **4**, 133 (1984).
114. D. R. Corbin, R. D. Farlee and G. D. Stucky, *Inorg. Chem.*, **23**, 2922 (1984).
115. M. W. Anderson, J. Klinowski and X. Liu, *J. Chem. Soc., Chem, Commun.*, 1596 (1984).
116. R. M. Dessau and G. T. Kerr, *Zeolites*, **4**, 315 (1984).
117. F. M. M. Geurts, A. P. M. Kentgens and W. S. Veeman, *Chem. Phys. Lett.*, **120**, 206 (1985).
118. M. C. Cruickshank, L. S. D. Glasser, S. A. I. Barri and I. J. F. Poplett, *J. Chem. Soc., Chem. Commun.*, 23 (1986).
119. Y. V. Shulepov, A. S. Litovchenko, A. A. Melnikov, V. Y. Proshko and V. V. Kulik, *J. Magn. Reson.*, **53**, 178 (1983).
120. S. Schramm and E. Oldfield, *J. Chem. Soc., Chem. Commun.*, 980 (1982).
121. I. P. Appleyard, R. K. Harris and F. R. Fitch, *Zeolites*, **6**, 428 (1986).
122. K. F. M. G. J. Scholle and W. S. Veeman, *Zeolites*, **5**, 118 (1985).
123. H. K. C. Timken, G. L. Turner, J. P. Gilson, L. B. Welsh and E. Oldfield, *J. Am. Chem. Soc.*, **108**, 7231 (1986).
124. H. K. C. Timken, N. Janes, G. L. Turner, S. L. Lambert, L. B. Welsh and E. Oldfield, *J. Am. Chem. Soc.*, **108**, 7236 (1986).
125. R. E. Grim, *Clay Mineralogy*, 2nd ed., McGraw-Hill, New York, 1968.
126. R. M. Barrer, *Zeolites and Clay Minerals*, Academic Press, London, 1978.
127. H. Strunz and C. Tennyson, *Mineralogische Tabellen*, 5. Aufl., Academische Verlagsgesellschafft Geest & Portig, Leipzig, 1970.
128. N. C. M. Alma, G. R. Hays, A. V. Samoson and E. T. Lippmaa, *Anal. Chem.*, **56**, 729 (1984).

129. P. Barron, P. Slade and R. L. Frost, *J. Phys. Chem.*, **89**, 3880 (1985).
130. R. A. Kinsey, R. J. Kirkpatrick, J. Hower, K. A. Smith and E. Oldfield, *Am. Mineral.*, **70**, 537 (1985).
131. J. Sanz and J. M. Seratosa, *J. Am. Chem. Soc.*, **106**, 4790 (1984).
132. P. F. Barron, R. L. Frost, J. O. Skjemstad and A. Koppi, *Nature (London)*, **302**, 49 (1983).
133. J. G. Thompson, *Clays Clay Miner.*, **33**, 173 (1985).
134. R. L. Frost and P. Barron, *J. Phys. Chem.*, **88**, 6206 (1984).
135. P. F. Barron, M. A. Wilson, A. S. Campbell and R. L. Frost, *Nature (London)*, **299**, 616 (1982).
136. T. J. Pinnavaia, S. D. Landau, M. Tzou, I. Johnson and M. Lipsicas, *J. Am. Chem. Soc.*, **107**, 7222 (1985).
137. S. Komarneni, C. A. Fyfe and G. J. Kennedy, *Clays Clay Miner.*, **34**, 99 (1986).
138. M. Mägi, A. Samoson, M. Tarmak, G. Engelhardt and E. Lippmaa, *Dokl. Akad. Nauk SSSR*, **281**, 1169 (1981).
139. D. Plee, F. Borg. L. Gatineau and J. J. Fripiat, *J. Am. Chem. Soc.*, **107**, 2362 (1985).
140. M. Lipsicas, R. H. Raythatha, T. J. Pinnavaia, I. D. Johnson, R. F. Giese, Jr, P. M. Constanzo and J. L. Robert, *Nature (London)*, **309**, 604 (1984).
141. J. Sanz, J. M. Serratosa and W. E. E. Stone, *J. Mol. Struct.*, **141**, 269 (1986).
142. J. G. Thompson, *Clays Clay Miner.*, **32**, 233 (1984).
143. P. A. Diddams, J. M. Thomas, W. Jones, J. A. Ballantine and H. A. Purnell, *J. Chem. Soc., Chem. Commun.*, 1340 (1984).
144. D. Müller, I. Grunze, E. Hallas and G. Ladwig, *Z. Anorg. Allg. Chem.*, **500**, 80 (1983).
145. C. S. Blackwell and R. L. Patton, *J. Phys. Chem.*, **88**, 6135 (1984).
146. D. Müller, E. Jahn, G. Ladwig and U. Haubenreisser, *Zeolites*, **5**, 53 (1985).
147. D. Müller, E. Jahn, G. Ladwig and U. Haubenreisser, *Chem. Phys. Lett.*, **109**, 332 (1984).
148. N. S. Kotsarenko, V. M. Mastikhin, I. L. Mudrakovski and V. P. Schmachkova, *React. Kinet. Catal. Lett.*, **30**, 375 (1986).
149. E. M. Flanigen, B. M. Lok, R. L. Patton and S. T. Wilson, *Pure Appl. Chem.*, **58**, 1351 (1986).
150. D. W. Sindorf and G. E. Maciel, *J. Phys. Chem.*, **86**, 5208 (1982).
151. D. W. Sindorf and G. E. Maciel, *J. Am. Chem. Soc.*, **103**, 4263 (1981).
152. D. W. Sindorf and G. E. Maciel, *J. Phys. Chem.*, **87**, 5516 (1983).
153. D. W. Sindorf and G. E. Maciel, *J. Am. Chem. Soc.*, **105**, (1983).
154. C. A. Fyfe, G. C. Gobbi and G. J. Kennedy, *J. Phys. Chem.*, **89**, 277 (1985).
155. C. S. John, N. C. M. Alma and G. R. Hays, *Appl. Gatal.*, **6**, 341 (1983).
156. M. McMillan, J. S. Brinen and G. L. Haller, *J. Catal.*, **97**, 243 (1986).
157. L. B. Welsh, J. P. Gilson and M. J. Gattuso, *J. Appl. Catal.*, **15**, 327 (1985).
158. C. E. Bronnimann, I, Chuang, B. L. Hawkins and G. E. Maciel, *J. Am. Chem. Soc.*, **109**, 1562 (1987).
159. J. R. Schlup and R. W. Vaughan, *J. Catal.*, **99**, 304 (1986).
160. D. Hoebbel, G. Garzo, E. Engelhardt and A. Vargha, *Z. Anorg. Allg. Chem.*, **494**, 31 (1982).
161. G. Engelhardt, D. Hoebbel, M. Tarmak, A. Samoson and E. Lippmaa, *Z. Anorg. Allg. Chem.*, **484**, 22 (1982).
162. R. K. Harris, C. T. G. Knight and W. E. Hull, *ACS Symp. Seri.*, No. 194, 79 (1982), and references cited herein.
163. D Hoebbel and W. Wieker, *Z. Anorg. Allg. Chem.*, **384**, 43 (1971).
164. G. Boxhoorn, O. Sudmeijer and P. H. G. van Kasteren, *J. Chem. Soc., Chem. Commun.*, 1416 (1983).

165. R. K. Harris and C. T. G. Knight, *J. Chem. Soc., Faraday Trans. 2*, **79**, 1525 (1983).
166. R. K. Harris, M. J. O'Connor, E. H. Curzon and O. W. Howarth, *J. Magn. Reson.*, **57**, 115 (1984).
167. G. Engelhardt and D. Hoebbel, *J. Chem. Soc., Chem. Commun.*, 514 (1984).
168. E. J. J. Groenen, A. G. T. G. Kortbeek, M. McKay and O. Sudmeijer, *Zeolites*, **6**, 403 (1986).
169. C. T. G. Knight, R. J. Kirkpatrick and E. Oldfield, *J. Am Chem. Soc*, **108**, 30 (1986).
170. C. J. Creswell, R. K. Harris and P. J. Jageland, *J. Chem. Soc., Chem. Commun.*, 1261 (1984).
171. C. T. G. Knight, R. J. Kirkpatrick and E. Oldfield, *J. Chem. Soc., Chem. Commun.*, 66 (1986).
172. A. V. McCormick, A. T. Bell and C. J. Radtke. In Y. Murakami, A. Iijima and J. W. Ward (eds), *Proceedings of the 7th International Zeolites Conference*, Kodansha, Tokyo, 1986, p. 247.
173. C. T. G. Knight, R. J. Kirkpatrick and E. Oldfield, *J. Am. Chem. Soc.* **109**, 1652 (1987).
174. S. Komarneni, R. Rustum, C. A. Fyfe and G. J. Kennedy, *J. Am. Ceram. Soc.*, **68**, C243 (1985).
175. S. Komarneni, R. Rustum, C. A. Fyfe, G. J. Kennedy and H. Strobl, *J. Am. Ceram. Soc.*, **69**, C42 (1986).
176. G. Engelhardt, B. Fahlke, M. Mägi and E. Lippmaa, *Zeolites*, **3**, 292 (1983).
177. G. Engelhardt, B. Fahlke, M. Mägi and E. Lippmaa, *Zeolites*, **5**, 49 (1985).
178. K. F. M. G. J. Scholle, W. S. Veeman, P. Frenken and G. P. M. van der Velden, *Appl, Catal.*, **17**, 233 (1985).
179. D. Michel, A. Germanus, D. Scheller and B. Thomas, *Z. Phys. Chem. (Leipzig)*, **262**, 113 (1981).
180. D. Michel, A. Germanus and H. Pfeifer, *J. Chem. Soc., Faraday Trans. 1*, **78**, 237 (1982).
181. I. Junger, W. Meiler and H. Pfeifer, *Zeolites*, **2**, 310 (1982).
182. C. A. Fyfe, G. T. Kokotailo, J. D. Graham, C. Browning, G. C. Gobbi, M. Hyland, G. J. Kennedy and C. T. Deschutter, *J. Am. Chem. Soc.*, **108**, 522 (1986).
183. H. Lechert amd H. W. Henneke, *ACS Symp. Ser.*, No. 40, 53 (1977).
184. H. Lechert, *Catal. Rev. Sci. Eng.*, **14**, 1 (1976).
185. T. Tokuhiro, L. E. Iton and E. M. Peterson, *J. Chem. Phys.*, **78**, 7473 (1983).
186. W. D. Basler, *J. Chem. Phys.*, **82**, 5297 (1985).
187. D. Freude, U. Lohse, H. Pfeifer, W. Schirmer, H. Schmiedel and H. Stach, *Z. Phys. Chem. (Leipzig)*, **255**, 443 (1974).
188. G. W. West, *Zeolites*, **1**, 150 (1981).
189. P. Chu, B. C. Gerstein, J. Nunan and K. Klier, *J. Phys. Chem.*, **91**, 3588 (1987).
190. G. A. H. Tijink, R. Janssen and W. S. Veeman, *J. Am. Chem. Soc.*, **109**, 7301 (1987).
191. T. Ito and J. Fraissard, *J. Chem. Phys.*, **76**, 5225 (1982).
192. L. C. de Menorval and J. P. Fraissard, *J. Chem. Soc., Faraday Trans. 1*, **78**, 403 (1982).
193. J. A. Ripmeester, *J. Magn. Reson.*, **56**, 247 (1984).
194. J. A. Ripmeester, *J. Am. Chem. Soc.*, **104**, 209 (1982).
195. T. Ito, L. C. de Menorval, E. Guerrier and J. P. Fraissard, *Chem. Phys. Lett.*, **111**, 271 (1974).
196. J. Demarquay and J. Fraissard, *Chem. Phys. Lett.*, **136**, 314 (1987).
197. E. G. Derouane and J. B. Nagy, *Chem. Phys. Lett.*, **137**, 341 (1987).
198. T. Ito and J. Fraissard, *J. Chim. Phys.*, **83**, 441 (1986).
199. T. Ito and J. Fraissard, *J. Chem. Soc., Faraday Trans. 1*, **83**, 451 (1987).

200. L. C. de Menorval, J. P. Fraissard and T. Ito, *J. Chem. Soc., Faraday Trans. 1*, **78**, 403 (1982).
201. E. W. Scharpf, R. W. Crecely, B. C. Gates and C. Dybowski, *J. Phys. Chem.*, **90**, 9 (1986).
202. D. Freude, M. Hunger and H. Pfeifer, *Chem. Phys. Lett.*, **91**, 307 (1982).
203. H. Pfeifer, D. Freude and M. Hunger, *Zeolites*, **5**, 274 (1985).
204. D. Freude, M. Hunger, H. Pfeifer, G. Scheler, J. Hoffman and W. Schmitz, *Chem. Phys. Lett.*, **105**, 427 (1984).
205. D. Freude, M. Hunger, H. Pfeifer and W. Schwieger, *Chem. Phys. Lett.*, **128**, 62 (1986).
206. K. F. M. G. J. Scholle, W. S. Veeman, J. G. Post and J. H. C. van Hooff, *Zeolites*, **3**, 214 (1984).
207. K. F. M. G. J. Scholle, A. P. M. Kentgens, W. S. Veeman, P. Frenken and G. P. M. van der Velden, *J. Phys. Chem.*, **88**, 5 (1984).
208. J. A. Ripmeester, *J. Am. Chem. Soc.*, **105**, 2925 (1983).
209. W. P. Rothwell, W. X. Shen and J. H. Lunsford, *J. Am. Chem. Soc.*, **106**, 2452 (1984).
210. J. H. Lunsford, W. P. Rothwell and W. Shen, *J. Am. Chem. Soc.*, **107**, 1540 (1985).
211. P. D. Majors and P. Ellis, *J. Am. Chem. Soc.*, **109**, 1648 (1987).
212. P. D. Majors, T. E. Raidy and P. Ellis, *J. Am. Chem. Soc.*, **108**, 8123 (1986).
213. L. Balthusis, J. S. Frye and G. E. Maciel, *J. Am. Chem. Soc.*, **109**, 40 (1987).
214. H. Pfeifer. In P. Diehl, E. Fluck and R. Kosfeld (eds), *NMR: Basic Principles and Progress* Vol. 7, Springer, Berlin, 1972, p. 53.
215. M. G. Munowitz and R. Griffin, *J. Chem. Phys.*, **76**, 2848 (1982).
216. C. Schimidt, S. Wefing, B. Blümich and H. W. Spiess, *Chem. Phys. Lett.*, **130**, 84 (1986).
217. H. Pfeifer, W. Meiler and D. Deiniger, *Annu. Rep. NMR Spectros.* **15**, 291 (1983).
218. A. Germanus, J. Kärger, H. Pfeifer, N. N. Samulevic and S. P. Zdanov, *Zeolites*, **5**, 91 (1985).
219. J. Kärger and H. Pfeifer, *Zeolites*, **7**, 90 (1987).
220. R. Eckman and A. J. Vega, *J. Am. Chem. Soc.*, **105**, 4841 (1983).
221. R. Eckman and A. J. Vega, *J. Phys. Chem.*, **90**, 4679 (1986).
222. J. B. Nagy, E. G. Derouane, H. A. Resing and G. R. Miller, *J. Phys. Chem.*, **87**, 883 (1983).
223. B. Zibrowius, M. Bülow and H. Pfeifer, *Chem. Phys. Lett.*, **120**, 420 (1985).
224. Z. Luz and A. J. Vega, *J. Phys. Chem.*, **90**, 4903 (1986).
225. C. E. Bronnimann and G. E. Maciel, *J. Am. Chem. Soc.*, **108**, 7154 (1986).
226. Z. Luz and A. J. Vega, *J. Phys. Chem.*, **91**, 374 (1987).
227. P. H. Kasai and P. M. Jones *J. Mol. Catal.*, **27**, 81 (1984).
228. H. E. Gotlieb and Z. Luz, *J. Magn. Reson.*, **54**, 257 (1983).
229. C. F. Tirendi, G. A. Mills and C. Dybowski, *J. Phys. Chem.*, **88**, 5765 (1984).
230. P. J. Toscano and T. J. Marks, *J. Am. Chem. Soc.*, **107**, 653 (1985).
231. P. Wang, J. P. Ansermet, S. L. Rudaz, A. Wang, S. Shore, C. P. Slichter and J. H. Sinfelt, *Science*, **234**, 35 (1986).
232. C. P. Slichter, *Annu. Rev. Phys. Chem.*, **37**, 25 (1986).
233. C. D. Makowka, C. P. Slichter and J. H. Sinfelt, *Phys. Rev. B*, **31**, 5663 (1985).
234. P. Wang, C. P. Slichter and J. H. Sinfelt, *J. Phys. Chem.*, **89**, 3606 (1985).
235. I. D. Gay, *J. Magn. Reson.*, **58**, 413 (1984).
236. T. M. Duncan, J. T. Yates, Jr, and R. W. Vaughan, *J. Chem. Phys.*, **71**, 3129 (1979); **73**, 975 (1980).
237. R. K. Shoemaker and T. M. Apple, *J. Phys. Chem.*, **89**, 3185 (1985).
238. A. Michael, D. Michel and H. Pfeifer, *Chem. Phys. Lett.*, **123**, 117 (1986).
239. T. M. Duncan, P. Winslow and A. T. Bell, *Chem. Phys. Lett.*, **102**, 163 (1983).
240. T. M. Duncan, J. A. Reimer, P. Winslow, and A. T. Bell, *J. Catal.*, **95**, 305 (1985).

Analytical NMR
Edited by L. D. Field and S. Sternhell
© 1989 John Wiley & Sons Ltd

Chapter 6

Biological Applications of NMR

P. W. Kuchel

Department of Biochemistry, University of Sydney, Sydney, 2006, N.S.W., Australia

1 BACKGROUND AND SCOPE

This is a review of aspects of NMR spectroscopy which involve identifying and estimating the amounts and concentrations of solutes in biological material. The emphasis is on techniques, so biochemical results are discussed in the context of how they are obtained and not, in general, their importance to the understanding of living systems, although of course this understanding is the motivation for carrying out biological NMR measurements. The earliest reports of NMR spectra of cells were concerned with erythrocytes and their water and electrolyte status[1,2]. These studies, involving low-resolution ^{1}H NMR spectra, were followed by many others investigating the physical nature of 'biological water'[3]. The first high-resolution NMR spectra of cells, reported in 1973, were of the ^{31}P nucleus in rabbit erythrocytes[4]. These simple spectra, containing only about seven resonances, permitted the estimation of intracellular pH, and monitoring of changes in the concentrations of 2,3-bisphosphoglycerate (DPG), adenosine triphosphate (ATP) and orthophosphate (Pi) over a time course of many hours. Curiously, these studies were not extensively followed up by the original authors, but the utility of ^{31}P NMR spectroscopy as a non-invasive means of studying the chemistry of living systems was rapidly recognized by many others[5–7]. The burgeoning literature in the area of ^{31}P NMR of biological systems has been progressively reviewed in recent years[8–15].

Direct NMR observation of other nuclei in cell systems has become popular because the procedure is non-invasive, and may be highly specific. The spectroscopic analyses permit the definition of the chemical and physical microenvironment of cells. A good example is the estimate of intracellular microviscosity made using the longitudinal relaxation times of ^{13}C-containing metabolites. The microviscosity of the very highly concentrated glycerol, or its glycosyl derivative, in each of two species of alga[16] has been measured this way, as has the microviscosity inside human erythrocytes using ^{13}C-labelled glycine and glutathione[17,18]. The use of ^{13}C NMR to study metabolic processes is gaining favour amongst biochemists both with the advent of high-field spectrometers with greater sensitivity and with the ready availability of ^{13}C-labelled substrates[19,20]. The general principles governing the use of ^{13}C NMR for studying biological samples are similar to, but in general less complicated than, those relating to ^{1}H NMR.

The ^{1}H atom is the most abundant in living systems and it has the most NMR-receptive stable nucleus, but the ^{1}H NMR spectrum of cells is an 'embarrassment of riches' in terms of spectral lines[10]. Important technical advances that permit the acquisition of useful ^{1}H NMR spectra from a range of aqueous samples have been made. The information content of the ^{1}H spectrum is potentially the greatest of all nuclei and no isotope enrichment is required, so experimental procedures relating to it will be described first. In the next section the most important technical problem associated with ^{1}H NMR is dealt with, and then other matters relating to the ^{1}H nucleus are left until Section 9.

2   ELIMINATING THE ^{1}H$_2$O SIGNAL FROM ^{1}H NMR SPECTRA

2.1   General

The most significant instrumental problem with ^{1}H NMR is coping with the dynamic range of the signals. It is often necessary to observe solutes of concentration well below 1 mM, yet in most cases there is a potentially interfering signal much more intense than this, specifically, the ca 110 M ^{1}H$_2$O protons of water solvent, ca 30 M total concentration of protein protons in cells, and ca 1 M protons from residual ^{1}H$_2$O if ^{2}H$_2$O is used as the solvent (Section 2.2.1). Thus, the ratio of the ^{1}H$_2$O signal to that of most metabolites is ca 10^5:1. Even in modern NMR spectrometers the peak intensities in spectra acquired sequentially vary by ca 1%, so that difference spectra cannot be used for isolating a small signal next to one >100 times its size[21]. To emphasize this point, we can rearrange the equation for the shape of a Lorentzian line[22,23] to obtain an expression for the relative signal $[g(v)/g(v_0)]$ at the frequency v offset from the centre frequency v_0 of the water line, i.e.

$$\frac{g(v)}{g(v_0)} = \frac{1}{\left(\dfrac{2\Delta v}{\Delta v_{1/2}}\right)^2 + 1} \tag{1}$$

where $g(v_0)$ is the peak amplitude, $\Delta v = v_0 - v$ and $\Delta v_{1/2}$ (Hz) is the peak width of water at half-height. From equation 1 we calculate that at a position 160 Hz from the centre of the line, the signal is 10^{-5} that of the peak amplitude; for a typical ^{1}H$_2$O linewidth of ca 4 Hz in a cell suspension, this position corresponds to 1.6 ppm at 400 MHz.

Although many NMR spectrometers now have analogue-to-digital converters (digitizers) that are 16-bit, thus implying a dynamic range of $1:2^{16} = 1:6.6 \times 10^4$, it is still advantageous to reduce the size of the ^{1}H$_2$O signal before it reaches the digitizer. There are basically three categories of procedures for this: (a) chemical methods, (b) solvent saturation and (c) selective non-excitation of ^{1}H$_2$O.

2.2   Chemical ^{1}H$_2$O signal suppression

2.2.1   Replacement by ^{2}H$_2$O

The most obvious way to reduce the ^{1}H$_2$O resonance in cell spectra is to wash the cells centrifugally in ^{2}H$_2$O medium. This has the additional advantage of providing an internal field/frequency lock. This was used in our first studies of erythrocytes[24] and continues to be a useful approach. The rate of glycolysis in human erythrocytes appears not to be affected by this treatment, but solvent isotope effects are manifest in the kinetics of some erythrocyte enzymes, e.g. prolidase (E.C. 3.4.13.9)[25], γ-glutamyl amino acid cyclotransferase (E.C. 2.3.2.4)[26,27] and the oxidative part of the pentose phosphate pathway[28].

The dependence of pK_a values of ionizable groups on the $^2H_2O/^1H_2O$ ratio[29] and the uncertainty of the actual meaning of p^2H cast doubt on the precise relevance of studies carried out in ^2H_2O to the situation *in vivo*. pH electrodes that have been equilibrated in ^1H_2O do not record the correct $p^2H = -\log_{10}[^2H_3O^+]$ when immersed in ^2H_2O, but read ca 0.4 pH units lower than the true p^2H[21,30]. Some authors add 0.4 to the indicated pH and report it as p^2H and others use the uncorrected meter reading. The latter is preferable since it is virtually impossible to obtain a biological sample in which all the ^1H_2O has been exchanged for ^2H_2O. In addition, because the pK_a of a deuterium dissociating reaction in ^2H_2O is also increased by ca 0.4 units, the pH dependence of any binding phenomenon is more clearly indicated if the uncorrected pH meter reading is used[21].

It may be of interest to observe exchangeable protons in a macromolecule or metabolite, in which case the exchange of ^1H_2O for ^2H_2O is undesirable[31]. This fact, coupled with the wish to avoid kinetic isotope effects, necessitates additional methods for ^1H_2O signal suppression.

2.2.2 Exploiting fast chemical exchange

In the *water attenuation by T_2 relaxation* (WATR) experiment[32], the T_2 of ^1H_2O protons is selectively reduced by adding a special solute. The ^1H NMR spectrum is recorded using the Carr–Purcell–Meiboom–Gill (CPMG) $[\pi/2_x-(\tau-\pi_y-\tau)_n-\text{aquire}]$[33] sequence, or a modification of the 2D J-resolved sequence[34] or 2D COSY[32]. By making the time between the $\pi/2$ pulse and acquisition ($2n\tau$) sufficiently long, the water resonance may be completely eliminated from the spectrum. Phase modulation of homonuclear coupled multiplets (coupling constant J; Section 9.3.3) is suppressed by using τ values that are small relative to $1/J$.

Several classes of compound which have labile protons can be used as ^1H_2O suppression agents[32,34,35]. The rate of $^1H^+$ exchange between NH_4^+ and water is slow on the NMR time-scale at pH $\approx$ 3, it is intermediate at pH $\approx$ 6.5 and fast at pH $\geq$ 8.5[36]. At pH 6.5 the value of $1/T_2$ is maximized and is proportional to the NH_4^+ concentration, at least in the sub-molar range[32]. The 360 MHz ^1H spectrum in Figure 1 shows the dramatic reduction of ^1H_2O signal achievable with this technique. Equally dramatic results have been obtained using hydroxylamine (0.25 M)[34]. This exchange is the basis of the previously unexplained 'anomalous frequency dependence' of the T_2 of ^1H_2O[37].

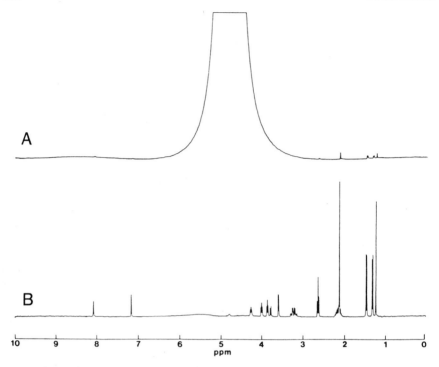

Fig. 1. ^1H_2O ^1H NMR peak suppression by chemical exchange. Spectrum A was obtained using a single-pulse method and spectrum B using the CPMG sequence, for a pH 6.5 solution containing 1 mM alanine, histidine, threonine and methionine in 0.5 M NH_4Cl in 99% ^1H_2O–1% ^2H_2O. Experimental conditions: A, 4° flip angle, 560 transients; B, $\tau = 0.003$ s, 670 180° pulses, 96 transients; line broadening factor, 0.3 Hz. From Ref. 32, with permission

Human urine normally contains ca 200 mM urea. A prominent resonance from the amide protons is evident at $\delta = 5.25$ ppm in the ^1H NMR spectrum of this fluid. At 400 MHz neutralized urine shows a broad peak ($\Delta v_{1/2} \approx 50$ Hz); this feature, interestingly, is not due to 'intermediate' exchange[23] of the amide protons with water but is the result of ^{14}N–^1H scalar coupling giving a 1:1:1 triplet which is partially decoupled because of fast ^{14}N quadrupolar relaxation[35,38]. The slow exchange (rate constant < 1 s^{-1}) of urea protons at neutral pH is faster, however, at pH ≈ 3; the observed T_2 value of ^1H_2O is a minimum at pH 3.2 and is proportional to the urea concentration[35,39]. ^1H_2O signal attenuation factors > 4000 can be achieved with urea concentrations of ca 2 M; high concentrations, however, can impair ^2H_2O signal locking because of broadening of that resonance.

The fact that no sample pretreatment, other than adding a solute, is required is of great analytical appeal. However, the method cannot be used with intact-cell samples or with biological fluids in which the activity of enzymes may need to be assayed.

2.2.3 Shift reagents

In principle, a paramagnetic reagent which preferentially complexes but exchanges rapidly with water could be used to shift the water resonance away from regions of particular interest in a spectrum. The hopes for p-sulphonatophenyl-porphyrinylterbium(III), which produces a 1 ppm high-frequency shift of ^1H_2O at a concentration of 20 mg ml^{-1}[40], seem not to have been realized for biological systems.

2.3 ^1H_2O radiofrequency irradiation

2.3.1 ^1H_2O nuclear magnetic saturation

The equilibrium nuclear magnetization of ^1H_2O can be reduced to zero by irradiating the sample with a selective radiofrequency (r.f.) field[41,42]. The appeal of this method is that it can be combined with many of the other solvent suppression techniques described here. A major disadvantage is that many protons in biomolecules are in rapid exchange with water so that if the water nuclei are saturated then so too are the exchanging ones by saturation transfer[40,43]. If the ^1H_2O saturation time is kept short and is applied just before the observation pulse, good suppression can still be achieved; some of this effect may be due to inhomogeneous nutation of the ^1H_2O magnetization. This sequence permits the observation of protons that exchange with solvent on a time scale longer than the pulse duration[44,45].

2.3.2 WEFT, IRSE and SE

The *water elimination Fourier transform* (WEFT) procedure is an old one[46,47] that can be used for detecting non-exchanging ^1H resonances near the water signal, usually in ^2H_2O media; or it can be used to suppress the ^1H_2O signal. The method exploits the differences between the T_1 of water (ca 2 s at 400 MHz in cells) and other molecules. A non-selective π pulse inverts all the magnetization of the sample and a delay τ is chosen during which all spins relax in the z-direction. The net ^1H_2O magnetization in free relaxation passes from being negative through zero to its positive equilibrium value. If τ is chosen so that a $\pi/2$ observe pulse is applied when the water magnetization is zero, then no ^1H_2O signal will appear in the spectrum. A homospoil pulse may be applied during the τ interval to eliminate transverse magnetization which may result from an imperfect π pulse[48].

The *inversion–recovery spin–echo* sequence (IRSE)[37] is the WEFT sequence in which the $\pi/2$ observe pulse has been replaced by the Hahn spin–echo sequence[24,49], i.e. $\pi-\tau_1-\pi/2-\tau_2-\pi-\tau_2$-acquire. Because water protons are in rapid exchange with some of those on haemoglobin in erythrocytes, the T_1 and T_2 values of the ^1H_2O are short compared with the protons of intracellular metabolites. Thus, with $\tau_1 = \tau_2 = 60$ ms, adequate ^1H_2O suppression in ^2H_2O-washed erythrocytes is obtained[50,51]. An even simpler approach, which is viable only with high-field, high-sensitivity spectrometers, is to use the spin-echo pulse sequence with long τ_2 values; values of the order of 240 ms can be used. This exploits the short T_2^* (apparent T_2) of ^1H_2O in erythrocytes, and presumably other cells, that occurs as a result of the previously mentioned exchange[37] (Section 2.2.2).

The T_2^* of ^1H_2O in erythrocytes decreases with increasing magnetic field strength, which is the opposite of that expected for a dipolar dependence[22]. The explanation is that, at lower magnetic field strengths, exchange with macromolecules may be fast ($\tau_c\Delta\omega \ll 1$)[23,52] where τ_c is the exchange lifetime and $\Delta\omega$ is the frequency difference between the exchanging species. However, at higher field strengths $\Delta\omega$ increases so that the intermediate exchange domain is entered ($\tau_c\Delta\omega \approx 1$). In this situation the two peaks of the exchanging system form a broad singlet[23,52]; such is the case with many of the haemoglobin protons and water.

2.4   Selective non-excitation of ^1H_2O

The ^1H_2O resonance intensity can be reduced in spectra by avoiding the excitation of water. There are basically four classes of methods to achieve this: (a) correlation spectroscopy, (b) soft-pulse and (c) hard-pulse sequences and (d) selective excitation of protons by polarization transfer from another nucleus.

2.4.1   Continuous wave and correlation spectroscopy

The first of these procedures involves slowly sweeping the B_1 (r.f.) field (or equivalently, B_0) through only the spectral region of interest. However, a major drawback is that most spectrometers now operate only in the pulse mode. One of the first ^1H NMR studies of metabolism in bacteria[53] used correlation spectroscopy[54,55]. In this method the B_1 field is swept rapidly over a frequency range that avoids ^1H_2O; the rapid sweep causes 'wiggles' to be associated with each peak in the spectrum. A separate signal is obtained from a compound with a single resonance or a reference signal is generated numerically. The experimental and reference data are subjected to mathematical deconvolution, which entails Fourier transformation and complex-number division to obtain a frequency domain spectrum; for the use of Fourier deconvolution, of exactly the same kind but in a different experimental context, see Ref. 56. A major disadvantage of the method is that experiments such as T_1 measurements are inconvenient to perform. Also, the theory of the data manipulation assumes a linear response from the spin system but at high r.f. fields and in some coupled systems this is not the case[40]. On the other hand decoupling of resonances very close to water, so-called 'under-water decoupling'[57], has been used for special problems; however, all of these problems could now be studied by using the methods described below.

2.4.2   Soft-pulse sequences

Redfield's[21,40,58,59] '214' pulse sequence achieves a relatively broad spectral region over which ^1H_2O is selectively not excited, i.e. the r.f. excitation has zero spectral density at the solvent resonance frequency. The technique has enjoyed much success in many laboratories, including our own[31], but it has been largely superseded by hard-pulse procedures because of their ease of implementation on modern spectrometers.

2.4.3   Hard-pulse sequences

It is possible to design 'hard' pulse sequences which nutate spins in some spectral regions by a large angle while not flipping others, such as ^1H_2O. Figure 2 shows the behaviour of magnetization vectors during three simple pulse sequences; these sequences are the so called 1-1[60], 1-$\bar{1}$ (also called 'jump and return' or JR)[61] and the 1-$\bar{2}$-1 [62] sequences. The digits indicate the relative durations of non-selective (hard) pulses and the over-bar denotes a 180° phase-shift of the pulse.

The search for pulse sequences which selectively do not excite water (or other spectral regions, e.g. extracellular $^{23}Na^+$ [63], Section 6) has been aided by the fact that for small nutation angles in an NMR experiment the excitation function in the frequency domain can be approximated by the Fourier transform of the pulse sequence in the time domain[64-66]. Hore[64,65] chose the function

$$S_n(\omega) = \sin^n(\omega\tau/2) \tag{2}$$

with n being a positive integer. All its derivatives, with respect to ω, up to the $(n-1)$th are zero at $\omega = 0$ (and at $\omega\tau = \pm 2\pi, \pm 4\pi, \ldots$). The inverse Fourier transform[67-69] of $S_n(\omega)$ is[65]

$$\sum_{k=0}^{k=n} (-1)^k \binom{n}{k} \delta(t + [k - n/2]\tau) \tag{3}$$

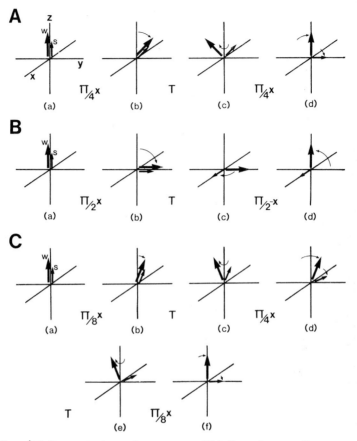

Fig. 2. Three ^1H_2O suppression pulse sequences. The Cartesian coordinate system is the rotating frame of reference, B_0 is in the z direction and w and s denote water and solute vectors, respectively. The operation of the r.f. pulse sequence on the vectors is as follows. The (A) 1–1 r.f. pulse sequence[60]; the 1–1 denotes two non-selective ('hard') pulses of the same phase. (a) At equilibrium w and s are pointing in the z direction then a $\pi/4x$ pulse is applied; (b) nutation of the vectors therefore takes place in the yz plane so that after the pulse the vectors lie at an angle of $\pi/4$ relative to the z axis; (c) during the time interval τ, since the Larmor frequency of s is set to being the same as that of the rotating frame, s remains in the yz plane whilst w precesses about the z axis into the $-yz$ subplane; (d) the second $\pi/4x$ pulse nutates w and s into the directions of the z and y axis, respectively, and signal acquisition begins. (B) The JR (or 1–Ī)[61] r.f. pulse sequence entails applying two $\pi/2$ pulses of appropriate phase, and it operates on w and s as follows: (a) to (b) the $\pi/2x$ pulse nutates w and s in the yz plane to the y axis (b) to (c) the Larmor frequency of w and the rotating frame frequency are set to be the same, so w remains in the y direction while s precesses in time τ into the x axis; (c) to (d) the $\pi/2 - x$ pulse nutates w into the z axis and does not affect s. (C) The 1–2–1 r.f. pulse sequence involves three pulses of the same phase, and of the relative durations indicated by the digits[62]. Based on (A) and (B), the operation of the sequence on w and s should be obvious. As a result of the sequence at the start of signal acquisition w lies along the z axis and s lies along the y axis

This function consists of $n + 1$ equally spaced (separation τ s) delta functions[67] with alternating sign and amplitude given by the binomial coefficient $\binom{n}{k}$. Equation 3 thus suggests a family of pulse sequences $1-\bar{1}$, $1-\bar{2}-1$, $1-\bar{3}-3-\bar{1}$, $1-\bar{4}\frac{1}{N}6\frac{1}{N}\bar{4}\frac{1}{N}1$,... For solvent suppression, the cumulative flip angle is chosen to be $90°$ for the solutes.

The key approximations in this derivation of pulse sequences are that relative nutation angles, rather than the amplitudes of the pulses, are given by $\binom{n}{k}$ and that the pulses of finite length approximate δ functions[65,70]. For reasonably strong (i.e. short) pulses and small frequency offsets the above argument is not too misleading. Figure 3 shows the excitation-power spectrum defined by various powers of the sine function (Equation 2).

The corresponding cosine functions,

$$C_n(\omega) = \cos^n(\omega\tau/2) \tag{4}$$

lead to the sequences without phase shifts, $1-1$, $1-2-1$, $1-3-3-1$, ..., which have $(n-1)$th-order nulls at $\omega\tau = \pm\pi, \pm3\pi, \ldots$; e.g. see Figure 2C. An even binomial sequence of order n is defined as a series of constant-phase pulses with flip angles

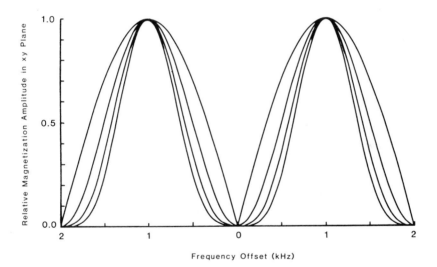

Fig. 3. Excitation envelopes for the odd binomial pulse sequences $1-\bar{1}$, $1-\bar{2}-1$, $1-\bar{3}-3-\bar{1}$ and $1-\bar{4}-6-\bar{4}-1$ as a function of transmitter frequency; these sequences correspond to the modulus of $\sin^n(\omega\tau/2)$, with $n = 1$, 2, 3 and 4, respectively. The curves all refer to $\tau = 500$ μs, rotating frame frequency $= 5$ kHz and nutation angle of the first pulse equal to $4.5°$, $22.5°$, $11.25°$ and $5.625°$, respectively. Note: these curves are those of the trigonometric function, whilst those calculated from the pulse sequences using rotation operators show a decidedly flatter region around the excitation maximum[65]; in other words, the experimental outcome from the pulse sequences is more favourable than indicated here

in the ratio of the coefficients of the binomial series for order n, separated by equal delays τ. The corresponding odd sequence is that with phases of alternate pulses different by 180°. In the limit of large n an even binomial sequence approaches a Gaussian function so that its excitation spectrum consists of a Gaussian-shape repeated every $1/\tau$ Hz. Similarly, an odd binomial sequence produces an excitation envelope with a null every $1/\tau$ Hz.

It should be apparent that there is an infinite variety of possible pulse sequences but an important constraint is their total time; this must be kept short to avoid chemical exchange or relaxation effects. Much effort continues to be expended on devising experiments to test the sensitivity of the sequences to instrumental imperfections, such as $\mathbf{B}_0$ inhomogeneity, $\mathbf{B}_1$ inhomogeneity and non-ideal r.f. pulse shape and phase shifts. For example, the $1-\bar{1}$(JR) method would suppress the ^1H_2O signal to only one hundredth of the original signal were the phase of the $-x$ channel in error by 0.6°, all other things being perfect[65].

Another problem arises at the signal processing stage. In many cases a phase gradient is imposed across the spectrum; e.g. with the $1-\bar{3}-3-\bar{1}$ sequence the spectrum has a phase shift which varies linearly with frequency, and a 180° step at the transmitter frequency[65]. However, both of these effects can be automatically corrected with appropriate modifications to the spectrometer software.

An even binomial sequence, with an extra antiphase pulse at its mid-point, gives an excitation spectrum which is flat except for nulls at offsets 0 and $\pm n/\tau$[71]. These sequences are called OBTUSE (*offset binomial tailored for uniform spectral excitation*), and two useful sequences have the form $\bar{1}-3-(4 + 4 + \varepsilon_1)-3-\bar{1}\frac{1}{N}\overset{\lrcorner}{\varepsilon}_2$ and $\bar{1}-5-\overline{10}-(16 + 16 + \varepsilon_1)-\overline{10}-5-\overset{\lrcorner}{\varepsilon}_2-\bar{1}$, where the centre bracketed terms denote composite 90° pulses with a 'trimming' adjustment ε_1, and $\overset{\lrcorner}{\varepsilon}_2$ is a 'phase-correcting' pulse applied with a phase shift of 90° (denoted by the $\lrcorner$) relative to the other pulses. High degrees of ^1H_2O suppression are achieved by optimizing both ε_1 and $\overset{\lrcorner}{\varepsilon}_2$[71].

The above pulse sequences are described as either symmetric or asymmetric, depending on the phase relationship of pulses about the centre of the sequence. A family of antisymmetric sequences has also proved to be effective for solvent suppression[66]. An antisymmetric sequence has the first null of the excitation region exactly on resonance, whereas symmetric sequences have this offset from the carrier frequency. The sequences $1-3-\bar{3}-\bar{1}$ and $3-5-15-\overline{15}-\bar{5}-\bar{3}$ were derived from the Fourier transformation of the repeating square function shown in Figure 4. The Fourier transform of this function is[66]

$$S(t) \propto \sum_{n=0}^{\infty} \frac{1}{2n + 1} \left[\delta\left(t - \frac{2n + 1}{2l} \right) - \delta\left(t + \frac{2n + 1}{2l} \right) \right] \tag{5}$$

where l is the frequency width (Hz) of the excitation function. Approximations to $S(t)$ can be obtained by taking various values of n. Hence the above pulse

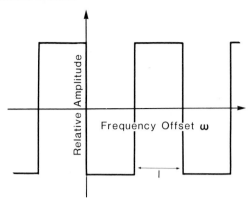

Fig. 4. Square-wave frequency domain function. The Fourier transform of this function gives a time domain r.f. pulse sequence which in principle excites this frequency domain

sequences have actual pulse widths that are calculated from the 'effective' pulse width (P_{eff}) for a resonance at $\pm l/2$, i.e.

$$P_{eff} = 2\left[\sum_{k=1}^{n} (-1)^{k+1} P_k \right] \tag{6}$$

where P_k is the pulse width and n is the number of terms taken in approximating equation 5.

The latest generation of solvent suppression pulse sequences is called NERO (*non-linear excitation with rejection on-resonance*)[72]; the sequences generate a broad null and excitation plateaux, and obviate the need for 'strong' spectral phase corrections after Fourier transformation.

The only general way of predicting the complete frequency response of a given r.f. pulse sequence is by numerical integration of the Bloch equations[62,72]. This is very computer-time consuming, but Levitt and Roberts[72] used an analysis called 'coherent averaging theory' to trace the response of spins in the neighbourhood of zero offset. Using the constraints that (a) the excitation must be zero on resonance, (b) the excitation must have at least a cubic dependence on offset frequency, thus defining a broad null, and (c) the offset dependence near the carrier frequency is insensitive to small errors in $\mathbf{B}_1$, a computer search was carried out for a sequence with a small number of pulses and delays. The resulting sequence NERO-1 has the form

$$120°-\tau_1-\overline{115°}-\tau_2-\overline{115°}-2\tau_3-115°-\tau_2-115°-\tau_1-\overline{120°}-\tau_r$$

where the τ_i values are given in Ref 72. The sequence is clearly asymmetric and the presence of large flip angles and variable τ values betrays its radical difference from the binomial sequences.

The extent of solvent suppression achieved in preliminary studies appears to be no better than for the other sequences[72] but, having now defined a more general strategy for searching for sequences, new and better ones will doubtless follow.

2.4.4 Heteronuclear polarization transfer

It is often desirable to study macromolecules, or metabolizing cells, in ^1H_2O and thereby avoid using ^2H_2O, which may give rise to isotope effects or exchange. An alternative to the above procedure uses the high sensitivity of ^1H NMR but the NMR selectivity of ^{13}C-enriched compounds. Thus the nuclear excitation of the compound takes place at a frequency far removed from the ^1H_2O frequency.

The 'reverse' polarization transfer pulse sequences, such as reverse-INEPT[73,74], reverse-DEPT[75] and reverse-POMMIE[76] have the potential to achieve ^1H_2O suppression in ^1H NMR spectra. Indeed, the metabolism of ^{13}C-labelled glucose by mouse liver-cell extracts has been studied by this means[77]. Preliminary results obtained using the reverse-POMMIE sequence with [1-^{13}C]ethanol (20 mM) in ^1H_2O suggest that the procedure may be useful for studying metabolic reactions *in vivo* without the '^1H_2O problem'[78].

2.4.5 Homonuclear polarization transfer (HPT)

The HPT experiment[79] is, in effect, a homonuclear modification of the re-focused-INEPT experiment[80]. The HPT pulse sequence discriminates between coupled and uncoupled resonances by polarization transfer within a homonuclear spin system. Pulse alternation is used to cancel uncoupled resonances and constructively add signals from coupled spins. The pulse sequence is $\pi/2-\tau-\pi_y-\tau-\alpha_y-\tau-\pi_y-\beta_{-y}$-acquire, where the delay $\tau = 1/4J$, and the third pulse α alternates between 0° and 90°. The alternation of α causes the phase behaviour of the coupled spins to differ from uncoupled spins. The pulse β alternates between 90° and 0° and is a purging pulse used to remove phase distortions and to correct for imperfections in τ.

Although the sequence can successfully suppress solvent singlets[79], it does not avoid the dynamic range problem (Section 2.1) and relies on precise signal subtraction to achieve its selectivity.

3   CELL VOLUME

3.1   General

An accurate estimate of the concentration of an intracellular solute requires a knowledge of the intracellular water volume or, more precisely, the volume of water available to the solute. Several procedures utilizing NMR have recently been introduced to measure this volume. Because of their simplicity these methods may supplant the older ones that employ radiotracers (see, e.g., Ref. 81), or even the newer less direct method using laser light scattering[82].

3.2 ^{35}Cl/^{2}H method

The resonance of intracellular ^{35}Cl$^-$ ions in most cells is so broad as to be NMR invisible[83,84]. Therefore, the ^{35}Cl NMR signal intensity of a cell suspension is proportional to the extracellular water volume. The signal intensity from a cell-free solution in an identical sample geometry is proportional to the total sample volume detected by the NMR coils. Thus the ratio of the two intensities (f_{Cl}) gives the extracellular water volume (V_e) as a fraction of the total volume of the cell suspension, i.e.

$$f_{Cl} = \frac{V_e}{V_e + V_c} \tag{7}$$

where V_c is the total cell volume.

The ratio of the intensities (f_D) of the ^{2}H NMR signal of the solvent in the cell suspension and the cell-free extracellular fluid, separated by mild centrifugation of the cells, gives the water volume as a fraction of the total volume of the cell suspension. Thus,

$$f_D = \frac{V_e + V_i}{V_e + V_c} \tag{8}$$

where V_i is the intracellular water volume. Therefore, the intracellular water volume as a fraction (f_w) of the total cell volume is given by[84]

$$f_w = \frac{V_i}{V_c} = \frac{f_D - f_{Cl}}{1 - f_{Cl}} \tag{9}$$

For human erythrocytes in Hanks balanced salt solution, with 2% ^{2}H$_2$O, this method yielded a value of f_w of 0.734 ± 0.024[84].

A variation of this method involves using an impermeable shift reagent, such as 5–8 mM Dy^{3+}-diethylenetriaminepentaacetic acid (Dy^{3+}DTPA), to separate the resonances of intra- and extracellular ^{1}H$_2$O and thus obtain an estimate of the ratio of intra- to extracellular water volume without the need to centrifuge the cell suspension to obtain a supernatant. The chemical shift of the ^{2}H$_2$O resonance is measured in the pure medium (δ_i) and then in the medium containing the shift reagent(δ_e). In a cell suspension the chemical shift (δ_{obs}) of the ^2H_2O, which is rapidly exchanging across the cell membranes, will have a value intermediate between the two extremes, and dependent on the relative proportion of intra- and extracellular water. Thus[85],

$$f'_D = \frac{V_e}{V_e + V_i} = \frac{\delta_{obs} - \delta_i}{\delta_e - \delta_i} \tag{10}$$

and hence

$$f_w = \frac{V_i}{V_e} = \frac{(1/f'_D - 1)}{(1/f'_{Cl} - 1)} \tag{11}$$

By using this procedure it was shown[85] that at 37 °C in isotonic Ringer solution (pH 7.4), the fraction of rat kidney proximal tubules that is exchangeable-water space is 0.769.

3.3 $^{59}Co/^1H$ method

$^{59}Co(imidazole)_6{}^{3-}$ is impermeant to cell membranes and has been used in a 2 mM solution in a manner equivalent to $^{35}Cl^-$ above. The high receptivity of ^{59}Co (28% of ^1H), the lack of interfering intracellular signals and the closeness of its resonance frequency to ^{13}C and ^{23}Na give it some advantages over $^{35}Cl^-$ for cell water volume measurements.

Although the ^2H nucleus is much less receptive than ^{59}Co, ^2H_2O can be safely added to cells up to the 1% level without fear of affecting metabolism; but even potential isotope effects of ^2H_2O can be avoided. Thus, Ogawa et al.[86] measured the relative internal water space in a mitochrondrial suspension by using ^1H NMR. The ratio (R) of the resonance intensities of 0.25 M mannitol or sucrose, which are membrane impermeant, to that of ^1H_2O in the mitochondrial suspension is divided by the same ratio (r) measured from the supernatant[86]. Thus the intramitochondrial volume is calculated from

$$\frac{V_i}{V_e} = \frac{r}{R} - 1 \tag{12}$$

The osmotic effects of the high concentrations of mannitol or sucrose may preclude the use of this method in many cell systems.

$Co(CN)_6{}^{3-}$ is also membrane impermeant and has been used to obtain the ratio of intra- to extracellular volume of erythrocytes in order to calculate the intracellular $^{23}Na^+$ concentration[87]. However, because the authors did not measure the exchangeable-water volume (Section 3.2), their estimate of Na^+ concentration is expressed on the basis of 'per litre of cells', not cell water. Later work corrected this deficiency[88].

Cowan et al.[89] measured the chemical shift (δ_i) of ^1H_2O, using ^1H NMR, in centrifugally packed erythrocytes. They also made the measurement (δ_{Dy}) in buffer doped with DyPPP (dysprosium tripolyphosphate; see Section 6.1), and in a cell suspension in the latter medium (δ_{obs}). Thus, from the equivalent of equation 10 it was possible to obtain the ratio of extracellular ^1H_2O to the total water volume. However, this analysis does not yield the water volume inside the cells, so the estimate of $^{23}Na^+$ concentration in the erythrocytes is given as a function of *total* cell volume[89] (in fact, equation 2 in Ref. 89 is missing the term α_0 in the denominator of the expression for intracellular $[Na^+]$).

A potentially serious problem with the above method is that the shift reagent increases the magnetic susceptibility of the sample and thus shifts all spectral resonances, including that of ^1H_2O, to higher frequency. This results in an incorrect value for δ_{obs}. Further, the ligand state of haemoglobin can affect the magnetic susceptibility of a sample and thus alter the value of δ_i in different samples. Therefore, an internal chemical shift standard is needed to correct for this effect.

3.4 ^{31}P methods

3.4.1 ^{31}P NMR and dimethyl methylphosphonate

Since the cell volume may change during metabolism there is a clear advantage in having a means of continuously monitoring volume without the need for repeated cell sampling, or changing the spectrometer frequency. A method using ^{31}P NMR has recently been developed[90,91]; it is based on the observed cell-volume dependence of the chemical shift of dimethyl methylphosphonate [DMMP; $CH_3PO(OCH_3)_2$]. The method can be applied to high-haematocrit red cell suspensions and does not require shift reagents.

DMMP is a small, neutral, water-miscible molecule. In saline solutions the proton-decoupled ^{31}P NMR spectrum is a singlet at $\delta = 39.4$ ppm. Addition of whole erythrocytes to this solution causes the peak to be split and, as the average cell volume is decreased, by increasing the extracellular osmotic pressure the peak separation increases (Figure 5)[90]. The separation is linearly dependent on the mean cell haemoglobin concentration (*MCHC*), provided that haemoglobin is in the carbonmonoxy form[92]. In recent work the slope (*m*) of this calibration line was found to be the same for cells from all of nine donors investigated[92]. Therefore, if the MCHC of a cell sample is determined, by standard haematological means, at the beginning of a ^{31}P NMR time course

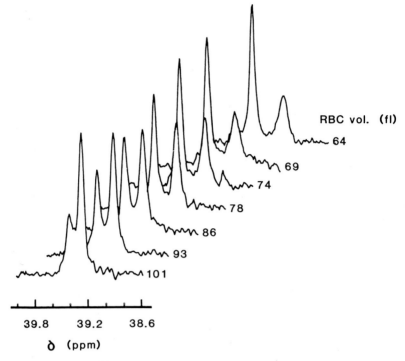

Fig. 5. Variation of the ^{31}P NMR (162 MHz) signal of DMMP with erythrocyte volume. Cells (haematocrit 0.875; 1.6 ml) were dispensed into a series of NMR tubes (10 mm o.d.); the extracellular osmotic pressures were then modified by adding various concentrations of NaCl (100 μl; 0–1.25 M). DMMP was added at a final concentration of 8 mM, then the ^{31}P NMR spectrum was recorded for each sample (8 transients, repetition time 60 s). The mean cell volume in each of the tubes was calculated from the measured haematocrit and cell count, and is shown on the right of the relevant spectrum. From Ref. 90, with permission

($MCHC_0$), the $MCHC$ at any subsequent time can be obtained from the ^{31}P spectrum by using the expression

$$MCHC = \frac{\Delta v - \Delta v_0}{m} + MCHC_0 \qquad (13)$$

where Δv_0 and Δv are the initial and subsequent DMMP peak separations (Hz), respectively. If the mean cell haemoglobin content (MCH) is also independently measured by using standard haematological techniques, then the mean cell volume (MCV) can be calculated using a rearrangement of equation 13:

$$MCV \text{ (fl)} = MCH \times 10^3 \bigg/ \left(\frac{\Delta v - \Delta v_0}{m} + MCHC_0 \right) \qquad (14)$$

This *MCV* is not V_i in equations 8-11 but, since the red cell membrane represents only ca 1% of the red cell volume[93], the value is close to V_c in equation 7. Further, 95% of human red cell protein is haemoglobin, so from its partial specific volume $(0.7546 \text{ ml g}^{-1})$[94] the necessary calculation can give an estimate of the intracellular water content.

Other features of the DMMP peak separation in suspensions of human red cells are that (a) the separation is independent of pH in the range 6.2-7.6; (b) it is independent of the haematocrit in the range 0.30-0.65; (c) it changes by -0.94 $\text{Hz} \, {}^{\circ}\text{C}^{-1}$ in the temperature range 25-65 °C; (d) it changes with DMMP concentration by 38 Hz/mol l^{-1} in the range 30-300 mM; and (e) it is not linearly dependent on *MCHC* if the ligand state of haemoglobin is varied such as by using methaemoglobin or partially oxygenated haemoglobin[92].

The physical basis of the DMMP peak separation in red cell suspensions is thought to be the different degrees of hydrogen bonding of water to the phosphoryl oxygen of this polar molecule inside and outside the cell[91,95]. Split ${}^{31}\text{P}$ peaks are also seen in the spectra of several 'phosphate analogues' in the presence of red cells, notably dimethylthiophosphate, thiophosphate, fluorophosphate, phosphite and hypophosphite. A number of these compounds do not titrate significantly in the physiological pH range[96,97] and a transmembrane pH difference therefore cannot account for the split peaks, as it does for a species such as methylphosphonate or Pi[98]. The claim that the effect is due to a transmembrane magnetic susceptibility difference[96,97] becomes untenable when it is noted that the peak separation differs amongst the solutes. In any case, the maximum theoretical susceptibility difference is insufficient to account for the observed resonance separations.

3.4.2 ${}^{31}\text{P}$ NMR estimation of intracellular water volume

We propose here a procedure akin to that in Section 3.2, but which does not require two different NMR-receptive nuclei[99]. The method obviates retuning the spectrometer, and has the additional advantage of involving a metabolically important nucleus with solutes whose nuclei do not resonate in the usual spectral window for metabolites.

To a cell suspension we add an isotonic solution (volume V_a) of a membrane-impermeant solute, such as pyrophosphate (PPi) or PPP, and a freely permeating solute TEP[100]. ${}^{31}\text{P}$ NMR spectra of the original solution and of the cell suspension are recorded; the ratio of the signal intensities of solutes is measured for each spectrum. By reasoning similar to that in the previous sections it can be shown that the ratio of the intracellular solute volume (V_i) to the extracellular volume (V_e), is given by

$$\frac{V_i}{V_e} = \frac{TEP_0}{PPP_0} \cdot \frac{PPP}{TEP} - 1 \tag{15}$$

where the three-letter variables denote the signal intensities and the subscript 0 refers to the original isotonic solution of solutes.

With erythrocytes the haematocrit (H_t) can be readily measured by microcapillary centrifugation and

$$H_t = \frac{V_c}{V_e + V_c} \tag{16}$$

where V_c is the cell-volume fraction and $V_e + V_c$ is the total volume of the cell suspension. Therefore, from equations 14 and 15 we obtain an expression for the ratio of intracellular volume to total cell volume that is akin to equation 11:

$$\frac{V_i}{V_c} = \left(\frac{TEP_0}{PPP_0} \cdot \frac{PPP}{TEP} - 1\right)\frac{(1 - H_t)}{H_t} \tag{17}$$

Alternatively, the haematocrit need not be measured by centrifugation. Instead of recording a ^{31}P NMR spectrum of the original TEP–PPP stock solution, we add the volume V_a of stock solution to buffer of volume equal to that of the original cell suspension. The ratio of the PPP signal intensities from the cell supernatant (PPP) and the new buffer mixture (PPP_B) is given by

$$\frac{PPP}{PPP_B} = \frac{1}{1 - H_t} \tag{18}$$

Therefore,

$$H_t = 1 - \frac{PPP}{PPP_B} \tag{19}$$

which can be substituted into equation 16, noting also that $TEP_0/PPP_0 = TEP_B/PPP_B$, to give an estimate of V_i/V_c.

4 pH

4.1 General

The concentration of free protons (or hydronium ions) lies in the nano- to micromolar range in most living systems; hence, their direct detection by ^{1}H NMR is impossible. However, by virtue of the fast exchange of protons between water and various bases, an NMR means exists for measuring pH. In fact, NMR is perhaps the most attractive of all methods for measuring intracellular pH[101].

The chemical shift of an NMR resonance of a titratable solute, expressed in terms of the chemical shifts of the acid (HA) and conjugate base (A^-) is[101–104]

$$\delta = \frac{[A^-]\,{}^b\delta + [HA]\,{}^a\delta}{[HA] + [A^-]} \tag{20}$$

where ${}^a\delta$ and ${}^b\delta$ are the chemical shifts of the acid and base, respectively, and the square brackets denote molar concentration. The Henderson–Hasselbalch equation may be written in the form

$$pH = pK'_a + \log_{10}\!\left(\frac{\delta - {}^a\delta}{{}^b\delta - \delta}\right) \tag{21}$$

where pK'_a is the apparent pK_a of the equilibrium reaction. Equation 20 may be fitted, by non-linear regression, to experimental data consisting of pH versus δ for a solute with a single ionization. This gives estimates of pK'_a, ${}^a\delta$, ${}^b\delta$ and the corresponding variances and correlation coefficients[98]. The method has been used with a range of NMR-receptive nuclei and in many different biological systems.

4.2 ^{31}P NMR methods

4.2.1   Commonly used probe molecules

Two important requirements of a pH-probe molecule are (a) a large difference in chemical shift between ${}^a\delta$ and ${}^b\delta$ (Equation 19) and (b) a value of pK'_a in the physiological pH range. Moon and Richards[4] were the first to use NMR for measuring pH; they measured the pH in erythrocytes by using the pH dependence of the chemical shift of Pi and also the difference in chemical shift between the 2-P and 3-P resonances of DPG. Many biological phosphate esters have pK'_a values in a range which is useful for pH studies[105]. However, these pK'_a values in some situations are profoundly influenced by the ionic and macromolecular composition of the medium. For example, increasing the ionic strength from 0.1 to 0.5 M with KCl in a 5 mM Pi solution, or in glucose 6-phosphate (G6P), displaces the titration curves so that the pK'_a values decrease by ca 0.38 and ca 0.30 units, respectively[106]. On the other hand, proteins, including bovine serum albumin (50 g l^{-1}) and a maize root tip protein, tend not to affect the titration curves of Pi and G6P even in the presence of $MgCl_2$[107]. The basic protein protamine does, however, greatly decrease the pK'_a of both of these solutes. The protamine effect on the titration curves is thought to operate via an electrostatic mechanism which alters the pK'_a, whereas K^+ and Mg^{2+} appear to exert specific binding effects that only alter ${}^a\delta$ and ${}^b\delta$[107].

In complex biological solutions the simple analysis that depends on equation 20 must be extended to incorporate additional ionizations. Because of the effects of Mg^{2+} on Pi and phosphate-ester chemical shifts, alternative metabolically inert probe molecules have been sought. A very successful one is methylphosphonate (MeP), which was first used to measure the intracellular pH of *Escherichia coli*[108]. The molecule has subsequently been used to measure the pH in human erythrocytes[98,109-111]; its NMR titration parameters in various media, together with those of three other important compounds, are given in Table 1.

Figure 6 shows a ^{31}P NMR spectrum of human erythrocytes obtained after brief incubation with MeP at 37 °C. The compound enters the cells via the anion-exchange protein band 3[98]. Note the high resonance frequency of MeP, which is well separated from the spectral region of most metabolites. Figures 7A and B show the titration curves for MeP and Pi, and Figure 7C shows the difference of the chemical shifts of these two solutes versus pH. The resonances of the two solutes move in opposite directions with a pH change. Since each solute has a different pK'_a the difference curve (Figure 7C) extends the range for measuring pH. Further, the use of the difference curve obviates the need for an absolute chemical shift reference and also compensates for the magnetic susceptibility difference which may arise between the intra- and extracellular compartments. Specifically, if the magnetic susceptibility of the medium used to construct the reference titration curves is different from that in the cells, an error in the pH estimate might otherwise occur[112].

The difference curve shown in Figure 7C was generated with the equation

$$pH = -\log_{10}[(B - (B^2 - 4AC)^{1/2})/2A] \qquad (22)$$

where

$$A = K_1 K_2(^a\delta_2 - {}^a\delta_1 - \Delta\delta); \quad B = K_1(^b\delta_2 - \Delta\delta - {}^a\delta_1) + K_2(^a\delta_2 - \Delta\delta - {}^b\delta_1);$$

$$C = {}^b\delta_2 - {}^b\delta_1 - \Delta\delta; \quad \Delta\delta = \delta_2 - \delta_1;$$

K_1 and $K_2 = 10^{pK_{a,1}}$ and $10^{pK_{a,2}}$, respectively; and subscript 1 denotes Pi and 2 denotes MeP.

Whereas the intracellular environment appears to influence the chemical shift of Pi more than it does MeP, ionic strength changes affect both species to comparable extents. The graphs of chemical shift versus ionic strength at pH 7 are slightly curved and have slopes of 1.212 and 0.829 ppm l mol^{-1} for Pi and MeP, respectively, at the physiological ionic strength 0.154 M[98].

A detailed error analysis of the MeP − Pi difference-curve procedure shows that the standard deviation (σ) of pH estimates in erythrocytes is $< \pm 0.03$ in the pH range 6.6–8.2[98]. Hence NMR can give precise estimates of intracellular pH, although on many occasions the claimed level of accuracy has not been based on rigorous error analysis.

TABLE 1
**Parameter estimates from regression of the modified Hender-
son–Hasselbalch equation on to ^{31}P NMR titration data**

Equation 20 was regressed on to titration data from ^{31}P NMR resonances in
erythrocyte lysates, modified Krebs hydrogencarbonate buffer (Krebs) and plasma/
additive solution (P/As). Values are $\pm$ approximate S.D.; n, number of data pairs; a
and b, chemical shift of the acid and base, respectively; MeP, methylphosphonate;
DPG, 2,3-bisphosphoglycerate. From Ref. 98, with permission.

Resonance	Medium	n	pK_a	$^a\delta$	$^b\delta$
MeP	Lysate	12	7.56 ± 0.02	25.131 ± 0.002	21.352 ± 0.055
MeP	Krebs	12	7.64 ± 0.04	25.160 ± 0.002	21.140 ± 0.178
MeP	P/As	11	7.52 ± 0.01	25.208 ± 0.003	21.259 ± 0.035
P_i	Lysate	12	6.85 ± 0.04	1.543 ± 0.071	3.670 ± 0.009
P_i	Krebs	12	6.59 ± 0.05	0.983 ± 0.066	3.441 ± 0.010
P_i	P/As	11	6.72 ± 0.01	1.039 ± 0.012	3.607 ± 0.004
DPG:3P	Lysate	12	7.32 ± 0.05	3.722 ± 0.018	4.811 ± 0.021
ATP-γ	Lysate	10	6.81 ± 0.15	-5.082 ± 0.113	-4.311 ± 0.010

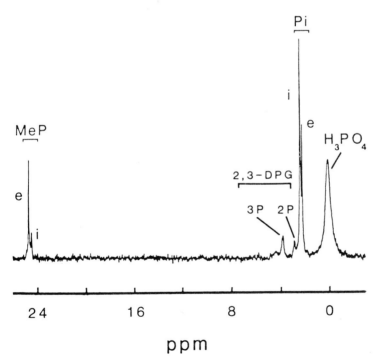

Fig. 6. ^{31}P NMR (162 MHz) spectrum of stored human erythrocytes after brief
incubation with MeP. The cell concentrate stored for 2 weeks had an extracellular pH of
6.64, and an intracellular pH of 6.70 based on the spectrum. Splitting of resonances from
MeP and Pi arose from the extracellular (e) and intracellular (i) pools. From Ref. 98, with
permission

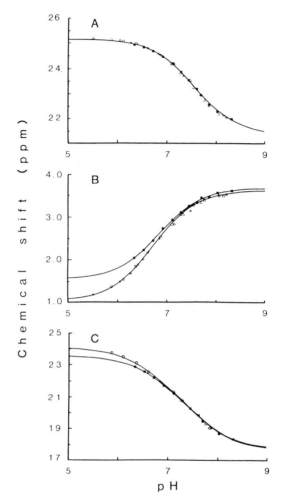

Fig. 7. pH dependence of the chemical shift of the [31] P NMR resonance of (A) MeP and (B) Pi in haemolysates (■), modified Krebs hydrogencarbonate buffer (△) and plasma/additive solution (○). The solid lines are regressions of equation 20 on to the data; details of the parameters are given in Table 1. Panel C shows the pH dependence of the difference in chemical shift ($\Delta\delta$) between the Pi and MeP resonances. For clarity the data for Krebs buffer have been omitted. The curves were calculated by substituting the parameters from Table 1 into equation 21. From Ref. 90, with permission

4.2.2 Other probe molecules

There are situations when Pi cannot be detected because of its low intracellular concentration, and MeP cannot be used because the particular membrane is impermeable to it. For example, MeP uptake by *E. coli* requires the induction of the glycerol transporter[113].

Phenylphosphonate freely enters bacteria by dissolution in the membrane. It has a ^{31}P NMR resonance that moves with pH changes up to 2.0 ppm between the extreme values of ca 13.8 and ca 11.8 ppm with a pK'_a of ca 7.2[114]. By using the difference between the titration curves of phenylphosphonate and Pi (as used for MeP/Pi, Section 4.2.1) the intracellular pH of *E. coli* was found to be 7.5 ± 0.05.

Because no Pi resonance appears in the ^{31}P NMR spectrum of freshly prepared perfused guinea-pig hearts, intracellular pH cannot be readily measured in this organ. 2-Deoxyglucose added to the perfusion medium, however, enters the myocardium and is rapidly phosphorylated via the hexokinase reaction. This product is (a) not a substrate for the next enzyme in the glycolytic pathway, glucose phosphate isomerase, (b) only slowly metabolized via the hexose monophosphate shunt, and in brain (but apparently not tested in myocardium) by several pathways including dephosphorylation by a hexose phosphatase[115], and (c) a glycogenolysis inhibitor[116]. Therefore, the 2-deoxyglucose 6-phosphate accumulates in the cells to levels that give a readily discernible resonance at −4.2 ppm with a maximum shift range between the acid and base forms similar to Pi[117].

The deoxyglucose method was also used to measure the cytoplasmic pH in maize root tips[118] and to show that Ehrlich ascites tumour cells regulate their intracellular pH, maintaining it at pH 7.1 when the extracellular value was 6.8–7.2[119].

The endogenous G6P resonance is sufficiently prominent in some biological circumstances to enable it to be used as a pH probe, although the precision of the estimate, when the method was used with plants, was not high (ca ± 0.2 units)[120].

Inosine monophosphate (IMP)[121] and fructose 1-phosphate[122] have also been employed as pH indicators, but their use has been limited to specific tissues which accumulate the compounds.

The chemical shift of the γ^{31}P of ATP in chromaffin granules was used to measure pH changes in these organelles. The pK'_a of the γP of ATP is ca 5, hence from the chemical shift of the resonance from intragranular ATP the pH was calculated to be 5.6 ± 0.1[123]. However a large number of ionic species affect the pK'_a of the γP of ATP (and analogues of it), thus making the establishment of a reference titration curve a difficult exercise[124].

4.2.3 Conclusion

The ^{31}P NMR method for measuring intracellular pH has been applied to a vast number of cellular systems from bacteria, plants and animal tissues and even in whole animals[15]. Care must be exercised in constructing the calibration titration curves, because ionic and macromolecular species that are found inside cells, but which may not be present in the standard solutions, may severely alter the pK'_a

values. Finally, the method of choice, and that subjected to the most detailed statistical analysis, appears to be the Pi/MeP method (Section 4.2.1).

4.3 ^{1}H NMR methods

The pH dependence of the chemical shifts of the H^2 and H^4 resonances of the histidine residues of haemoglobin allows the measurement of intracellular pH[24]. Standard titration curves are constructed from data obtained with a concentrated lysate. Great care is needed to avoid the magnetic susceptibility-induced shifts that arise from the effect of pH on the oxygen affinity of haemoglobin (the Bohr effect[110]). There are also direct effects of oxygen[125] and DPG binding[126] on the histidyl chemical shifts.

The chemical shift versus pH titration curves of the H^2 and $H^{4,5}$ of imidazole (5 mM) added to both haemolysates and saline solutions can be fitted by the modified Henderson–Hasselbalch equation (equation 20)[127]. However, different pK_a' values are obtained for the H^2 and $H^{4,5}$ resonances in the lysates, thus indicating the presence of molecular interactions which are different for each part of the molecule. Nevertheless, imidazole has been used to measure the intracellular pH of human erythrocytes; it slowly permeates the cell membrane and spin–echo spectra yield well resolved H^2 and $H^{4,5}$ resonances[127]. The maximum shift of the H^2 resonance is only 0.8 ppm with $pK_a' = 7.01$ at 37 °C; the corresponding values for $H^{4,5}$ are 0.44 ppm and $pK_a' = 6.99$[127]. Hence the method is not as precise as that described in Section 4.2; the authors nevertheless claimed a standard deviation of the pH estimates of ± 0.02 units, but this was not based on a complete statistical analysis.

Yoshizaki et al.[128] obtained 100 MHz ^{1}H NMR spectra of amphibian and mammalian muscle and detected resonances of, amongst other solutes, carnosine (β-alanylhistidine); they suggested that carnosine might be useful used as an intracellular pH probe. The limiting chemical shifts and pK_a' appropriate to equation 20 were determined for a solution containing 2 mM $MgCl_2$, which was added in order to mimic the intracellular concentration of Mg^{2+}, which is the most abundant of the intracellular divalent cations and which interacts with the dipeptide. The titration parameters were $^a\delta = 7.246$ ppm, $^b\delta = 6.906$ ppm and $pK_a' = 7.12$, so from the spectra of seven muscle samples the mean pH and standard error of the mean was 7.31 ± 0.05[129]. Earlier studies of frog muscle, using this technique[128,130,131], gave slightly lower estimates of the intracellular pH, but the accuracy of these results cannot be determined because $^a\delta$ and $^b\delta$ values were not reported for the titration.

Some muscle such as that from rat also contains the dipeptide anserine (β-alanyl-N-methylhistidine). In this case the chemical shift of the histidyl H^4 resonance versus pH, in a solution containing 2 mM $MgCl_2$ conformed to equation 20 with $pK_a' = 7.09$, $^a\delta = 7.253$ ppm and $^b\delta = 6.766$ ppm[129]. The authors emphasize the importance of an internal chemical shift reference to account for magnetic susceptibility-induced shifts[112,132,133]. A useful ^{1}H NMR internal shift reference in muscle is the methyl group of phosphocreatine, which resonates at 3.02 ppm[129].

'Dense granules' from blood platelets of reserpine-treated pigs contain high concentrations of 5-hydroxytryptamine (5HT; also called serotonin), histamine and ATP[134]. The aromatic region of the ^{1}H NMR spectrum contains resonances assigned to the above imidazoles and purine moieties. At the pH (ca 5.5) found inside the granules there appears to be no evidence of histamine binding to the nucleotides, so the pK'_a value is assumed not to be altered from that measured in dilute solutions. In dilute solutions the H^2 resonance of histamine had titration parameters of $^a\delta = 7.72$ ppm and $^b\delta = 8.65$ ppm, and the H^4 parameters were $^a\delta = 7.05$ ppm and $^b\delta = 7.40$ ppm[134]. The ^{1}H spectrum of the granules revealed histamine resonances at ca 7.83 and ca 7.08 ppm for the H^2 and H^4, respectively. In view of the expected pK'_a values near 6.9, these chemical shifts suggest an intragranular pH of ca 5.

The symmetrical dicarboxylic acids fumaric and maleic acid have 'first' pK'_a values of ca 3.8 and ca 5.8, respectively[135]. The extreme chemical shift values of the vinyl protons differ by less than 0.4 ppm, yet well resolved resonances are obtained from the acid and conjugate-base forms that exist in a suspension of phospholipid vesicles[135]. Compounds containing carboxyl groups, such as maleic and fumaric acid, could therefore be used to probe intracellular regions of low pH such as lysosomes; however, there do not appear to be any reports of their use with ^{1}H NMR. There is, however, one example, given below, in which ^{13}C malate was used as a pH probe.

4.4 ^{13}C NMR methods

There is a surprising paucity of information on intracellular pH as measured using titratable ^{13}C-labelled compounds. One report, however, describes a study of a crassulacean plant grown in an atmosphere containing ^{13}CO$_2$[136]. Malate ^{13}C-labelled in all four carbons was detected in a ^{13}C NMR spectrum of the plant. The pH dependence of the chemical shifts showed a maximum range of 24 ppm with the acid form at higher frequency. The pK'_a values of the two carboxyls were 3.5 and 5.1 ppm and 3.1 and 4.5 ppm at ionic strengths of 0.1 and 1.0 M, respectively[136]. The major compartment for malate is the plant vacuole, hence the pH estimated from the chemical shift of this solute refers to this organelle. The pH estimate made from a rather broad peak was 3.8-4.4. The wide pH range can be attributed to the heterogeneity of pH amongst vacuoles and the magnetic field inhomogeneity in the cells.

The 100 MHz ^{13}C NMR spectrum of a neutral solution of NaH^{13}CO$_3$ in water has two resonances separated by ca 3570 Hz; the larger peak is that of hydrogencarbonate and the smaller is $^{13}CO_2$. The pK'_a of the equilibrium reaction has a complicated dependence on a range of physical and chemical factors, but under 'cellular' conditions it is ca 6.1^{137}. Thus, from the ratio of peak intensities of hydrogencarbonate and CO_2 we were able to estimate the intracellular pH of a suspension of human erythrocytes[137].

4.5 ^{19}F NMR methods

The high NMR receptivity of ^{19}F relative to ^{31}P nuclei, and the lack of natural ^{19}F-containing metabolites, make the use of ^{19}F probe molecules appealing for pH determination. Trifluoroethylamine (TFE; $CF_3CH_2NH_2$) is relatively non-cytotoxic and the chemical shift of its ^{19}F resonance is strongly pH dependent; the maximum shift is 0.2 ppm per 0.1 pH unit near its pK'_a of 6.0, and pH can be accurately measured over the range 4–8. The ^{19}F spectrum of 4 mM TFE in a red cell suspension (haematocrit 0.33, in Tris–maleate buffer at 25 °C) showed a broad resonance from extracellular TFE ($\Delta v_{1/2} \approx 0.14$ ppm) and the intracellular resonance was not resolved from it when the extracellular pH was 7.2. At pH 8.0 a broad, unresolved shoulder on the high-frequency side of the extracellular TFE peak was evident[138]. The broadening was less evident at 4 °C and can be ascribed to fast transmembrane exchange of the solute. Deconvolution of the spectra was used to enhance the resolution and thus enable a pH estimate to be made; the precision of this estimate was claimed (with little justification) to be ± 0.02 units[138]. The accuracy of the method is improved by using an internal shift reference (Section 4.2.1) rather than the external trifluoroacetate in ^2H_2O that was used by Taylor et al.[138]. Internal trifluoroacetate (2.0 mM) has been used as a reference in subsequent work on lymphocytes[139].

A better ^{19}F pH probe, which does not exchange rapidly across the cell membrane, is D,L-2-amino-3,3-difluoro-2-methylpropionic acid (α-difluoromethylalanine). When human lymphocytes were incubated with the free amino acid, or its methyl ester, the free amino acid accumulated in the cells[140]. At neutral pH and 25–37 °C the methyl ester is taken up by the cells and endogenous esterase activity gives rise to intracellular concentrations of the free amino acid as high as 4.5 mM.

Both of the above compounds (the amino acid and its ester) give ^1H-decoupled ^{19}F NMR spectra which consist of an AB quartet with $J/\Delta v \approx 0.7$. The pK'_a of the α-amino group on the acid is 7.3 and on the methyl ester 5.1 (see Table 2). For both the ester and acid the spacing of the two centre lines of the quartet is given by

$$\Delta = [(v_A - v_B)^2 + J^2]^{1/2} - |J| \tag{23}$$

where v_A and v_B are the Larmor frequencies (Hz) of the two non-equivalent ^{19}F atoms and J is the ^{19}F–^{19}F coupling constant. The value of Δ increases

TABLE 2
pK'_a and maximum pH changes of fluorinated α-methylalanine derivatives[139]

Temperature, 25 °C; concentration of solutes, 1 mM.

Compound	pK'_a	$\Delta\delta$ per 0.1 pH unit
α-Trifluoromethylalanine methyl ester	<4	
α-Trifluoromethylalanine	5.9	0.12
Hexafluorovaline	6.3	0.06
α-Difluoromethylalanine methyl ester	5.1	0.14[b]
α-Difluoromethylalanine	7.3	0.09[b]
α-Monofluoromethylalanine methyl ester	6.4	0.08
α-Monofluoromethylalanine	8.5	0.11

[b] For the difluoromethylalanines $\Delta\delta$ is the change in frequency difference (in ppm) between the two centre peaks of the ^{19}F AB quartet.

markedly as the α-amino group is protonated and thereby serves as a pH indicator. This procedure of measuring peak separation circumvents the need for an internal chemical shift standard. The useful pH range is 6.6–7.8 for the acid and 4.3–5.9 for the ester. The maximum pH-dependent changes are given in Table 2.

^{19}F NMR spectra of tri- and monofluoromethylalanine have a single resonance; the titration parameters of the resonances are given in Table 2. The ^{19}F NMR signal of hexafluorovaline is a doublet, and its two resonances from the non-equivalent trifluoromethyl groups are 4 ppm apart at pH $\approx$ 7 with $J_{FF} =$ 8.5 Hz. The chemical shift of each resonance exhibits a different pH dependence but this compound is not useful as a pH probe because of the small pH-dependent shifts (Table 2).

The pK'_a values of both di- and trifluoromethylalanine are shifted in the alkaline direction when K^+ is added to a solution of these solutes; the effect 'saturates' at 130 mM. There is no similar effect with NaCl, though, and Mg^{2+} and Ca^{2+} have no significant effect in the 0–3 mM concentration range[139]. On the other hand, the shifts are sensitive to temperature; a temperature increase from 17 to 37 °C shifted the pK'_a of trifluoromethylalanine by 0.34 units while that of the difluoromethylalanine peak-difference changed by 0.8 units. Finally, both compounds showed separate intra- and extracellular ^{19}F NMR resonances in a lymphocyte suspension[139], isolated hepatocytes[141] and perfused frog skin[142]. Although the signal-to-noise ratio and resolution appear not to be as favourable as in the equivalent ^{31}P experiment (Section 4.2.1), similar pH values were estimated for frog skin using both methods.

The compound F-quene (Figure 8), based on the successful Ca^{2+} fluorescent probe quene 2 (which has CH_3O in place of the F), has a pK'_a of 6.7 and a chemical shift increase with an increase in pH from ca 7 to ca 12 ppm, to high frequency of nFBAPTA (see Section 5.2)[143]. This highly charged anion would

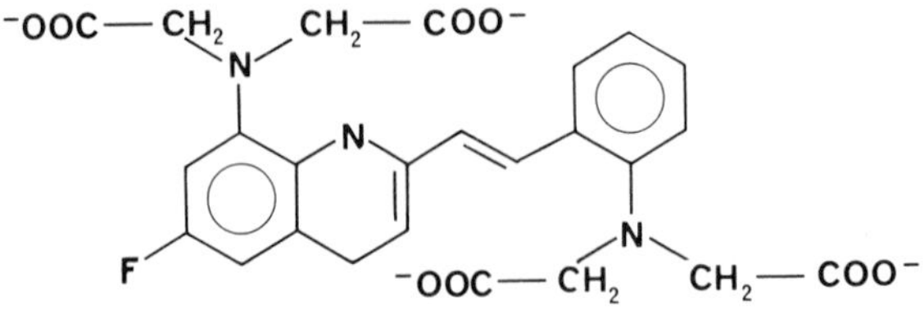

Fig. 8. The ^{19}F NMR pH-probe molecule F-quene

not cross most cell membranes but its tetramethyl ester may do so; subsequent intracellular esterase activity should release the pH-active species. Further work is required to establish whether this or related compounds will be useful ^{19}F NMR indicators of pH. However, some related compounds are valuable indicators of cell Ca^{2+}.

5   INTRACELLULAR Ca^{2+}

5.1   General

Calcium ions are key intracellular regulators of many enzymic and transport processes, e.g. the ions activate the K^+-leak channel[144] and inhibit the Na^+, K^+ATPase[145] of human erythrocytes. Calcium ions are involved in volume regulation of red cells from several animal species, including sheep[146], and high Ca^{2+} levels are implicated in the cell dehydration and membrane rigidity that occurs in sickle cell disease[147]. The concentration of free Ca^{2+} in most mammalian cells is in the submicromolar range in spite of plasma levels being around 5 mM.

The levels of Ca^{2+} in cells have been measured using a range of techniques, the most successful of which has been a fluorescent intracellular Ca^{2+} chelator[148–150]. The measurements, however, require very dilute cell suspensions.

Direct detection of intracellular free Ca^{2+} via the NMR-receptive nucleus, ^{43}Ca, is a hopeless prospect; ^{43}Ca has a natural abundance of 0.145, spin $I = 7/2$ and it has a receptivity relative to $^1H(D^P)$ of 9.26×10^{-6}[151]. Fortunately, a ^{19}F NMR-based method, that can in principle be applied to whole cells and tissues, has been described[152].

5.2 ^{19}F NMR measurement of Ca^{2+}

The symmetrically substituted difluoro derivatives of 1,2-bis(n-fluoro-o-amino-phenoxy)ethane-N,N,N',N'-tetraacetic acid (nFBAPTA, where n denotes the

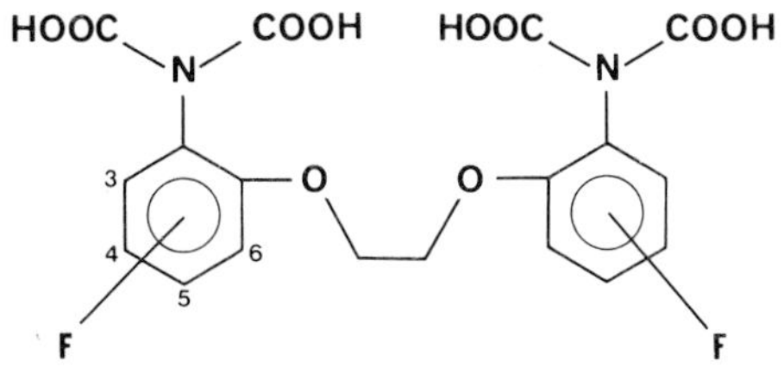

Fig. 9. The ^{19}F NMR Ca^{2+} indicator compound nFBAPTA

position of F substitution in each ring; see Figure 9) have ^{19}F NMR chemical shifts that are reasonably sensitive to chelation by divalent cations[152–154]. With 5FBAPTA the binding of Ca^{2+} shifts the ^{19}F resonance from 0.9 ppm, relative to 6-fluorotryptophan, by a maximum of 5.8 ppm. The extremes of chemical shift of 4FBAPTA are at 1.05 and 2.55 ppm[152].

The measurement of intracellular Ca^{2+} begins by incubating the cells with micromolar concentrations of the tetraacetoxymethyl ester of the reagent. The ester is only sparingly soluble in water so it is generally delivered in dimethyl sulphoxide (DMSO), for which it is desirable to keep a low (less than ca 20 mM) concentration within cells. The neutral species diffuses through the lipid of the plasma membrane of the cells, and cytoplasmic esterases 'unmask' the four carboxyl groups; this both prevents efflux from the cell and releases the free reagent, which is a polydentate chelator[143,154]. Intracellular concentrations of around 1 mM can be readily achieved in 45 min of incubation at 37 °C[143,155].

The Ca^{2+} binding stoichiometry with nFBAPTA is 1:1. In the case of the 3- and 5FBAPTA–Ca^{2+} complexes the ^{19}F NMR (94.1 MHz) spectra contain two separate resonances corresponding to free and bound Ca^{2+}, whereas with 4FBAPTA the two forms are in NMR fast exchange. In the latter case the single resonance is observed at a chemical shift between the frequencies of the free and bound forms, the shift position being weighted according to their relative proportions. This situation is akin to that with some $^1H^+$-titratable groups which are used as NMR probes of pH (Section 4).

The dissociation constant (K_d) of the Ca^{2+}–4FBAPTA complex, measured in citrate buffer, is ca 2.2×10^{-6} M. From a line-shape analysis of the ^{19}F NMR spectrum the dissociation rate constant for the complex was measured to be 9.8×10^2 s^{-1}[152]. The equilibrium dissociation constant of the Ca^{2+}–5FBAPTA complex, and its dissociation rate constant, were found to be 7.09×10^{-9} M and 5.7×10^2 s^{-1}, respectively[152]. The equilibrium constants and dissociation rate constants imply association rates that are approaching the diffusion limit (ca 0^9 mol^{-1} l s^{-1}); hence the chelators are capable of very rapid responses to changes in free Ca^{2+}.

Levy et al.[156] were critical of the properties of the nFBAPTA compounds because the K_d value of 5FBAPTA is only 5–10-fold lower than the free Ca^{2+}

levels and therefore the compound would not be saturated by these levels. Consequently, several new fluorinated chelators with high Ca^{2+} affinity were designed and synthesized. However, as stated above, the total Ca^{2+} in erythrocytes is vastly in excess of the free levels; hence a tight binding chelator may ultimately merely indicate the total amount of Ca^{2+} in the cell. High-affinity chelators might seriously influence metabolism by altering the 'available' intracellular Ca^{2+}.

The 'best' of the alternative Ca^{2+} indicators is 2-(2-amino-4-methyl-5-fluorophenoxy)-methyl-8-aminoquinidine-N,N',N',N'-tetraacetic acid (quin MF). Its K_d is ca 10 times lower than that of 5FBAPTA, and it has been used in experiments that yielded an estimate of 25–30 nM for free Ca^{2+} in erythrocytes[156].

In principle, the most appropriate indicator is one in which the complex is in fast exchange with the free species. Also, in order that an approximately linear relationship exists between the amount (and hence chemical shift) of complex and free ion concentration, the K_d of the complex should be much greater than the anticipated free ion concentrations. Unfortunately, the only fast exchange Ca^{2+}-indicator, 4FBAPTA, has a K_d that is much *less* than the intracellular free Ca^{2+} concentration, so its shift is insensitive to most intracellular Ca^{2+} levels.

The free Ca^{2+} concentration measured with 5FBAPTA is unaffected by (a) free Mg^{2+} at <10 mM, (b) pH in the range 6–8 or (c) minor cellular cations with

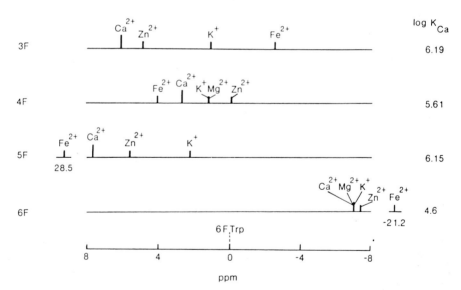

Fig. 10. ^{19}F NMR chemical shifts, from the external reference 6-fluorotryptophan, for complexes of nFBAPTA with M^{2+} ions at pH 7.1 and 37 °C. Shifts labelled K^+ denote positions of the resonances in a buffer containing 150 mM KCl and 10 mM HEPES at pH 7.1. From Ref. 152, with permission

high affinity for 5FBAPTA, since these complexes have different chemical shifts[152]. Figure 10 summarizes the relative affinities of the nFBAPTA derivatives for several divalent cations, and also emphasizes the markedly different chemical shifts of the various complexes.

In a recent series of experiments 5FBAPTA was used to estimate intra-erythrocyte free $[Ca^{2+}]$. To increase the sensitivity and precision of the method 20 mm NMR tubes were used, and concentrations were referenced to the integrated intensity of the signal from a known amount of added 6-fluorotryptophan[155]. The concentration was estimated as 61 ± 6.0 nM. The precision of this estimate is surprising when it is noted that the ^{19}F NMR linewidth of 5FBAPTA in the cell spectrum was typically 339 Hz and the signal-to-noise ratio of the spectra shown was ca $8:1$[155].

Ehrlich ascites cells labelled with 0.15 mm 5FBAPTA over a 2 h incubation period gave a ^{19}F NMR spectrum that contained three resonances, one from the free ligand, one from the Ca^{2+} complex and the third assigned to the Zn^{2+} complex. The concentration of the latter ion was inferred to be 0.4 mm[154]. In other work, both 5FBAPTA and F-quene (Section 4.5) were incorporated into perfused rat hearts, and from the ^{19}F NMR spectra the Ca^{2+} concentration and pH were estimated to be 690 nM and 7.1, respectively. The spectral accumulation time in these studies was ca 10–30 min each.

5.3 Conclusion

Although there is much hope for the success of ^{19}F-labelled indicators of free intracellular Ca^{2+}, their physical and chemical characteristics are as yet far from ideal. Because the chemical shift of a resonance is much more accurately defined than its intensity, an ideal Ca^{2+} indicator should (a) have a relatively low affinity for Ca^{2+}, (b) be in fast exchange with the complex, (c) be specific for Ca^{2+} and (d) be unperturbed in its binding by H^+, K^+, Mg^{2+} and other ions. None of these criteria are yet satisfactorily met by any of the above ligands.

6 Na$^+$ IN BIOLOGICAL SAMPLES

6.1 General

Most cells have a concentration disequilibrium of Na^+ ions across their plasma membrane[157]. The relatively high extracellular Na^+ concentration is maintained principally by Na^+, K^+-ATPase in the plasma membrane which competes with specific leak channels that allow Na^+ influx down its concentration gradient. Rapid discharge of the Na^+ gradient is the basis of nerve signal transmission and is the first event in nerve-mediated muscle contraction[157]. Altered Na^+ levels in cells are implicated in the aetiology of various diseases, e.g.

increased arteriolar smooth muscle Na^+, due to a circulating inhibitor of Na^+, K^+-ATPase, is found in essential hypertension[158] and [Na^+] changes occur during transformation of cells to the neoplastic state[159]. Amongst the many metabolic changes which reduce the value of stored blood for transfusion is elevated erythrocyte [Na^+][110].

It is clearly of vital importance to biochemists and clinicians to have estimates of intracellular [Na^+], and an important breakthrough in the quest for this information was an NMR means of distinguishing intra- from extracellular Na^+[153]. To a large extent Laszlo's 'prophecy'[160] of an explosive growth in the use of ^{23}Na NMR has already been fulfilled.

Like all the other alkali metals, ^{23}Na has a nuclear quadrupole moment; its spin $I = 3/2$, natural abundance of 100% and high receptivity, 525 times that of ^{13}C[151,160], ensure easy detection of $^{23}Na^+$ in cells and tissues.

6.2   Aqueous shift reagents

The ^{23}Na NMR resonance of $^{23}Na^+$ inside cells can be made anisochronous with that outside by using one of a variety of anionic shift reagents[153]. Thus, the above-mentioned transmembrane ionic disequilibrium can be measured continuously and non-invasively. Most success with shift reagents has been achieved with complexes of Dy^{3+}, the paramagnetic lanthanide which usually produces the largest of all hyperfine shifts[161,162]. While significant shifts of $^{23}Na^+$ frequency are obtained with $Dy(EDTA)^-$ (EDTA = ethylenediaminetetraacetate) even greater ones are evident with $Dy(DPA)_2^{3-}$ (DPA = dipicolinate) and $Dy(NTA)_2^{3-}$ (NTA = nitrilotriacetate), probably because of their large net negative charge which is available to bind Na^+[163]. The last compound is still used in biological work, and is prepared by mixing a $1:4$ molar ratio of Dy_2O_3 and NTA in buffer at neutral pH[163].

The stoichiometry of the interaction of Na^+ with $Dy(NTA)_2^{3-}$ is $1:1$ and the complex is labile on the NMR time scale. The shapes of the $^{23}Na^+$ lines are Lorentzian, corresponding to a weak interaction ($K_d > 0.1$ M) possibly involving ion pairing and the Na^+ being 'solvent separated'. The limiting chemical shift of $^{23}Na^+$ in the presence of a large excess (ca 5 times) of the reagent is > 10 ppm; this shift can be counteracted by the potently diamagnetic Lu^{3+} ion[164].

An example of the use of $Dy(NTA)_2^{3-}$ was to distinguish between intra- and extravesicular Na^+ in a suspension of synthetic unilamellar vesicles of egg lecithin. When the shift reagent was present on one side of the membrane two ^{23}Na NMR spectral resonances were seen. Dissipation of the Na^+ gradient by the ionophore gramicidin was able to be monitored over a period of ca 2.5 h[163].

Simultaneous with the report of the above-mentioned shift reagents was that of Dy^{3+} tripolyphosphate [PPP; $Dy(P_3O_{10})_2^{7-}$][165]. The reagent is prepared by mixing one volume of 0.1 M $DyCl_3$ and two volumes of freshly prepared 0.1 M pentasodium tripolyphosphate; it is added to cell suspensions to give a final extracellular concentration of ca 5 mM[166]. This is now the favoured reagent because (a) its low-frequency hyperfine shift is an order of magnitude greater than that of $Dy(NTA)_2^{3-}$; (b) it is relatively innocuous and appears not to react with most biomolecules or their complexes; (c) it appears to be chemically stable in biological solutions; (d) it does not significantly affect cell integrity; and (e) it appears not to cross cell membranes at significant rates, although recent work on human erythrocytes urges some caution in this regard[166].

Although the dissociation constant of the DyPPP complex has not been reported, some work suggests that small but significant amounts of free Dy^{3+} accumulate in cells during prolonged incubation[63,167]. Also, ^{31}P NMR spectra of erythrocytes show line broadening with time if Dy^{3+} is added to the extracellular medium[168]. Another potentially serious problem with DyPPP is that it forms a precipitating complex with Ca^{2+}; hence in principle it must be used with washed cells, and in the case of erythrocytes it should not be used in whole blood or Krebs hydrogencarbonate buffer[168]. Yet in one report, a non-spinning whole blood sample with 5 mM DyPPP gave a ^{23}Na NMR (81.8 MHz) spectrum that had two well resolved resonances separated by ca 6.5 ppm, $\Delta v_{1/2}$ values for the intra- and extracellular peaks were 41.5 and 81.8 Hz, respectively (Figure 11)[166]. Importantly, there was also no evidence of cell lysis in the sample.

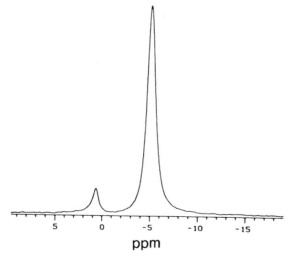

ppm

Fig. 11. ^{23}Na NMR (52.92 MHz) spectrum of fresh human erythrocytes in plasma with 5 mM DyPPP. NMR parameters: 20 mm NMR sample tube; 23 °C; 8192 data points; recycle time, 1.81 s; 100–200 transients per spectrum. From Ref. 166, with permission

In the perfusing Ringer solution used with a preparation of rabbit kidney proximal tubules the free Ca^{2+} concentration was detectably altered by DyPPP. In addition, the ATP content of the cells as measured by ^{31}P NMR was found to be reduced by 50% by added free PPP[169]. Therefore, care must be taken to prepare accurately a solution with the correct $1:2$ stoichiometry of Dy^{3+} and PPP.

6.3 Quantification of Na$^+$ concentration

6.3.1 [Na$^+$]$_i$ estimation

A comparison of the intensity of the $^{23}Na^+$ NMR resonances of extracellular ions (A_{out}) with that of a non-cellular control (A_0), containing the same concentration of Na^+ as is present in the extracellular medium, yields directly the fraction (S_{out}) of sample volume that is extracellular, i.e.

$$S_{out} = A_{out}/A_0 \tag{24}$$

The intensity (A_{in}) of the intracellular $^{23}Na^+$ resonance can be used to obtain the concentration of the ion per litre of cells[170]. Thus,

$$[Na^+]_i = \frac{A_{in}}{A_{out}}\left(\frac{S_{out}}{1 - S_{out}}\right)[Na^+]_{out} \tag{25}$$

If $[Na^+]_{out}$ is measured directly by flame photometry, or gravimetrically, then it is possible to obtain an estimate of the fraction of $^{23}Na^+$ that is 'NMR-visible'. A theoretical objection to the appropriateness of equation 24 is that is does not take into account the cell volume that is inaccessible to Na^+; in human erythrocytes this amounts to ca 30% of the cell volume (Section 3.2). More recent work by Kumar et al.[171] addressed this point in relation to kidney proximal tubule suspensions. They divided the right-hand side of equation 24 by $\alpha = V_i/V_c$ (Section 3.2) to yield an expression for the Na^+ concentration in the cell water. In contrast, though, in another investigation[172] the value of α for kidney tubules was not able to be defined, hence the Na^+ concentration was not able to be given precisely.

DyPPP has been used successfully to determine intra- and extracellular $[Na^+]$ in perfused rat hearts[166] and in dialysis microcapillary systems[168]. Some uraemic patients have elevated red cell $[Na^+]$ and ^{23}Na NMR-based estimates of the intracellular amounts of Na^+ agreed with those obtained with flame photometry. Statistical analysis revealed two groups of uraemic patients, one with a higher mean $[Na^+]$ of 8.5 mM than the other with 6.0 mM, while the control group had a mean concentration of 6.2 mM[170].

6.3.2 Invisible $^{23}Na^+$

The NMR lineshape of $^{23}Na^+$ that undergoes exchange between a free state in aqueous solution and a bound state of a slowly reorientating molecule is the superposition of two Lorentzian lines. The two components correspond to a fast and a slow relaxation representing 60% and 40% of the total signal intensity, respectively[173-176].

The biexponential relaxation occurs when the $-3/2 \rightarrow -1/2$ and $1/2 \rightarrow 3/2$ transitions of the ^{23}Na nuclei relax faster than the $-1/2 \rightarrow 1/2$ transition[173]. Under conditions that are favourable, both physically and numerically, the two lines can be resolved by numerical deconvolution of the spectra[176]. In biological systems $^{23}Na^+$ may bind tightly to macromolecules, thereby leading to more rapid relaxation of the already rapidly relaxing component, thus broadening it beyond detection. This effect could lead to a resonance with an apparent intensity of only 40% of the value expected for a given Na^+ concentration.

Much attention has been given to the NMR invisibility, or otherwise, of $^{23}Na^+$ in various cell systems. Values close to 40% visibility have been obtained for yeast cells[177] and *E.coli* cells[178] and in muscle, kidney, brain and liver[179]. On the other hand, 100% visibility of intracellular $^{23}Na^+$ has been noted for erythrocytes[180] and kidney proximal tubules[169].

The lineshape of the extracellular $^{23}Na^+$ resonance in an *E.coli* suspension was simulated by two Lorentzians with areas in the ratio 60:40, thus suggesting that Na^+ in the vicinity of the outside cell surface of the bacterium also experiences quadrupolar interactions[178]. The NMR visibility of the extracellular $^{23}Na^+$ is close to 100% despite the broad component, but the intracellular component has a single-Lorentzian lineshape corresponding to $45 \pm 5\%$ of the expected value (as noted above).

Much lower levels of NMR visibility of $^{23}Na^+$ were found in the prophase-arrested follicle-enclosed oocytes of the frog *Rana pipiens*; this 'low visibility' was attributed to both quadrupolar broadening and compartmental heterogeneity[181]. Progesterone induction of breakdown of the cell nucleus caused an increase in the visibility to 26%, and after fertilization the value rose to 70% of the maximum. The oocytes contain yolk platelets which in turn contain $^{23}Na^+$ which is NMR invisible. Further, it can be shown by using partially relaxed inversion–recovery spectra that there are also at least two main classes of cytosolic compartment[181], so the interpretation of the ^{23}Na NMR spectra of frog eggs is indeed complicated.

6.3.3   Alternatives to Dy^{3+}

Dextran–magnetite at an intravascular concentration of 1 μM has been used to give partial resolution of intra- and extracellular ^{23}Na NMR signals in the perfused hind limb of rats[182]. The potential advantage of this biologically inert complex is its apparent resistance to metabolism in contrast to the recently reported *in vivo* breakdown of PPP[183]. For isolated cells and tissues, however, it appears to hold no advantage over DyPPP.

6.4   Coherence transfer ^{23}Na NMR

An interesting development for the selective detection of intracellular ^{23}Na$^+$ is the use of double quantum filtering[184,185]. Interactions between the electric quadrupole moment of the ^{23}Na nucleus and fluctuating electric field gradients at Na$^+$ binding sites cause biexponential relaxation[173] (Section 6.3.2). Unless conditions of extreme narrowing hold, the $-3/2 \rightarrow -1/2$ and $1/2 \rightarrow 3/2$ ('outer') transitions of ^{23}Na relax faster than the $-1/2 \rightarrow 1/2$ ('inner') transition, even though only a small fraction of the Na$^+$ may be bound. Biexponentially relaxing ^{23}Na$^+$ can be passed through a state of double quantum coherence while that which monoexponentially relaxes cannot.

Distinguishing biexponential from monoexponential relaxation is difficult, especially if the two relaxation times do not differ by a factor of at least two. Also, rapidly relaxing transitions decay so quickly that the signal resulting from the outer transitions may be invisible (Section 6.3.2). Measurement of the two relaxation times is aided by using a one-dimensional multiple quantum filter[186]; the pulse sequence passes the spin system through a state of multiple quantum coherence and since the response of a spin 3/2 system can be quantitatively predicted it is possible to estimate the relaxation times[186]. By measuring ^{23}Na$^+$ peak height as a function of double quantum 'creation time' (one of the delays in the pulse sequence), the two relaxation times of ^{23}Na$^+$ in an albumin solution were measured, even though the relaxation rates differed by only a factor of four[184].

In a cell suspension for which only the intracellular ^{23}Na$^+$ relaxes biexponentially, selective detection of the intracellular ion may be achieved using a double quantum filter. To show that the intracellular sodium was indeed selectively detected, Pekar *et al.*[185] added DyPPP to a suspension of washed dog erythrocytes; Figure 12 shows that only the intracellular ^{23}Na$^+$ was detected by this elegant procedure.

Although it can be seen from Figure 12 that double-quantum filtering can in principle be used to detect selectively ^{23}Na$^+$ in cell components, in which the ^{23}Na$^+$ has different binding and relaxation behaviour, the 'sensitivity penalty' may be too large for many applications. On the other hand, the method has the

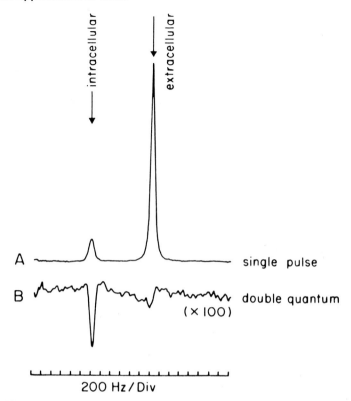

Fig. 12. ^{23}Na NMR (52.9 MHz) spectra of dog erythrocytes suspended in isotonic medium containing 12 mM DyPPP. (A) Conventional spectrum, obtained with 90° r.f. pulses, 64 acquisitions in 20 s. (B) double-quantum filtered spectrum obtained with the pulse sequence $\pi/2-\tau/2-\pi-\tau/2-\pi/2-\delta-\pi/2$-acquire with a 64-step cycle to detect double quantum coherence during the evolution period δ[186]; 3200 transients acquired in ca 17 min. The vertical scale was corrected for number of transitions. From Ref. 185, with permission

advantage of avoiding the use of shift reagents which may have adverse metabolic effects.

6.5 ^{19}F NMR indirect Na$^+$ detection

The value of indirect detection of Na$^+$, when the nucleus is easily detected directly by NMR, is the possibility of obtaining information on intracellular compartmentation based on the selective penetration of the 'indirect detector' into those compartments.

The requirements for a useful indirect NMR-based detector of Na$^+$ (or for that matter any other ion, see Section 5) are stringent[187]: (a) the spectroscopic parameters must be sensitive to Na^+ binding in the cytosolic concentration range 5–50 mM; (b) the ligand must be highly selective for Na^+ in the presence of intracellular K^+ (ca 130 mM), Mg^{2+} (0.5–5 mM) and free Ca^{2+} ($<10^{-6}$ M) and insensitive to pH in the range 6–8; (c) the indicator must be sufficiently polar to be retained in the cytosol, or whichever organelle is 'targetted'; (d) the free Na^+ concentration must not be perturbed by the detector molecule, which implies a low concentration of complex either by use of small amounts of the detector molecule on a low stability constant for the complex; and (e) there must be a minimal effect on cell metabolism.

Smith et al.[187] selected the cryptand–crown ether system for development as intracellular Na^+ indicators, because of its exquisite selectivity for Na^+ over K^+ and because of the facility with which derivatives can be synthesized perhaps to meet the above biological requirements.

The compounds have the basic structure shown in Figure 13. Of the many derivatives investigated, FCryp-1 appears to hold most promise.

FCryp-1 shows a 2.00 ppm ^{19}F NMR chemical shift between the free cryptand and Na^+ complex; its affinity for Na^+ is relatively low, $K_d = 50$ mM, and it is selective against K^+. At 376 MHz the ^{19}F NMR spectrum of the ligand–Na^+ complex is a single resonance indicating fast exchange binding. With pig lymphocytes, loaded with 0.21 mM FCryp-1 by incubating the cells in a medium containing 10 μM of the tetraacetoxymethyl derivative, the intracellular $[Na^+]$ was readily estimated from the ^{19}F NMR spectrum to be 13.8 mM. This value compared favourably with the result of direct ^{23}Na NMR measurements[187] and is consistent with the high degree of ^{23}Na NMR visibility of intracellular Na^+ in many cells (Section 6.3.2).

The ground work laid with the development of FCryp-1 is valuable and paves the way for interesting future investigations.

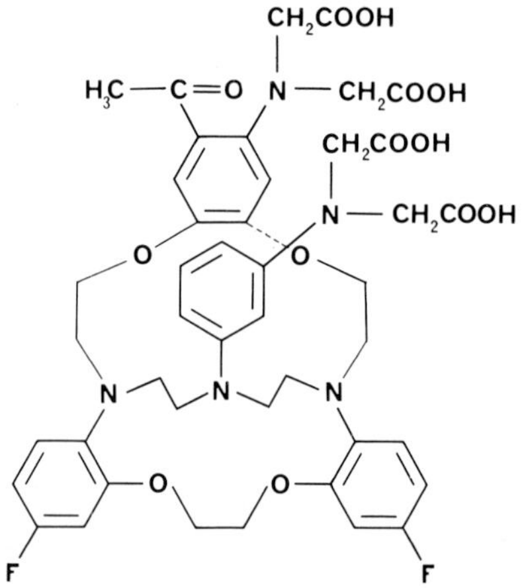

Fig. 13. The ^{19}F NMR detector molecule for Na^+, FCryp-1, made by Smith et al.[187]

7 ^{39}K IN BIOLOGICAL SAMPLES

K^+ presents more severe problems to the NMR spectroscopist than does Na^+ because ^{39}K is the least receptive of the alkali metals, although its natural abundance is 93.1% and spin $I = 3/2^{151}$. Fortunately, the intracellular K^+ concentration is often ca 20-fold greater than that of Na^+, so direct observation by NMR is possible in an acceptable spectral accumulation time.

Tripolyphosphate (PPP; Section 6.2) enhances the pseudo-contact shifting of $^{39}K^+$ resonances by Dy^{3+}. This was first exploited by Brophy et al.[167] in work on human erythrocytes and has subsequently been used to discriminate intra-from extracellular $^{39}K^+$ in perfused rat hearts[63], in further work on erythrocytes[188] and in vivo in rat muscle, liver, kidney and brain[189]. Impressively, the latter spectra were obtained in 5–10 min with a signal-to-noise ratio of ca 20:1, at 11.05 MHz ($B_0 = 5.8$ T). Quantification of K^+ concentrations was made by measuring the ratio of the ^{39}K to the ^{1}H signal in tissue, compared with that of a KCl standard in a tube extending beyond the r.f. coils, by using the 'in vivo quantitative' method of Thulborn and Ackerman[190]. For all in vivo studies changes in the $^{39}K^+$ signal were related to the initial control values using an external reference as an integration standard.

NMR visibility of $^{39}K^+$ is defined, like that for $^{23}Na^+$ (Section 6.3.2), as the percentage detection of the total ion as determined by flame photometry of a nitric acid tissue digest. For blood the visibility was 96%, but for muscle, liver, kidney and brain it was 47, 43, 45 and 62%, respectively[189]. These results are similar to those obtained for Na^+ (Section 6.3.2) and indicate intracellular binding to macromolecules with short correlation times, thus giving rise to biexponential relaxation.

Hence ^{39}K NMR is a viable direct procedure for measuring intracellular K^+ in a broad range of cells and tissues. The degree of visibility, which varies from tissue to tissue, however, is a serious problem in the quantification of the ion, although the altered relaxation behaviour of $^{39}K^+$ inside cells has the potential to give an insight into cytoplasmic organization.

8 INTRACELLULAR Mg^{2+}

8.1 General

Intracellular Mg^{2+} ions can influence a vast array of biochemical reactions; virtually all kinases and ATP-dependent reactions use the Mg complex of the nucleotide as their substrate. For example, hexokinase, the first enzyme in glycolysis, uses Mg-ATP as its substrate but it is inhibited by glucose 1,6-bisphosphate (Gl6P$_2$), DPG, Pi and Mg^{2+}-free ATP; thus Mg^{2+} levels have a profound effect on the rate of erythrocyte glucose metabolism[28]. Similar effects are found with cardiac muscle metabolism[191]. The regulatory effects of Mg^{2+} do not appear to be as finely tuned as those of Ca^{2+}, however, and the intracellular levels of the free ion are generally much higher. The concentrations of free Mg^{2+} in cells range from ca 0.4 mM in frog muscle to ca 6 mM in barnacle muscle[191].

The NMR-receptive nucleus ^{25}Mg is spin $I = 5/2$, of 10.13% natural abundance and receptivity 1.54 times that of ^{13}C[148]; therefore, with biological systems it is only possible to detect it indirectly with NMR spectroscopy.

8.2 ^{31}P NMR detection of Mg^{2+}

The ^{31}P NMR method for estimating free Mg^{2+} is based on the accurate measurement of the frequency difference between the αP and βP resonances in the spectrum of intracellular ATP. The chemical shifts of these resonances are sensitive to the state of divalent cation complexation, and since Mg^{2+} is usually the most predominant (see Section 5 for a comment on Zn^{2+} as a possible exception to this claim) of these cations the ^{31}P NMR spectrum allows the determination of the fraction of ATP that is complexed[192].

Fast exchange averaging of chemical shifts takes place between the free ATP and the ATP complex[153]; there is, however, some evidence of exchange broadening of the βP peak at high magnetic field strengths[193]. The ratio of the concentrations of free ATP to total ATP (Φ) is given by[153],

$$\Phi = \frac{[\text{ATP}]_{\text{free}}}{[\text{ATP}]_{\text{total}}} = \frac{\Delta\delta_{\alpha\beta}^{\text{cell}} - \Delta\delta_{\alpha\beta}^{\text{MgATP}}}{\Delta\delta_{\alpha\beta}^{\text{ATP}} - \Delta\delta_{\alpha\beta}^{\text{MgATP}}} \tag{26}$$

where $\Delta\delta_{\alpha\beta}^{\text{cell}}$ is the $\alpha - \beta$P chemical shift difference in the cell, $\Delta\delta_{\alpha\beta}^{\text{ATP}}$ is the value for an ATP solution under simulated intracellular conditions of pH and ions but in the absence of Mg^{2+} and $\Delta\delta_{\alpha\beta}^{\text{MgATP}}$ is the value in lysed cells to which is added an excess of Mg^{2+}. The free Mg^{2+} concentration is then given by

$$[\text{Mg}]_{\text{free}} = K_d^{\text{MgATP}}\left(\frac{1}{\Phi} - 1\right) \tag{27}$$

where K_d^{MgATP} is the MgATP dissociation and the apparent value of $50 \pm 10 \; \mu\text{M}$ is that usually used in the calculations[153]. Nevertheless, an exact value of K_d^{MgATP} is not strictly necessary if a range of Mg^{2+} control experiments is carried out[153].

As an example, in one series of experiments the $\alpha - \beta P$ chemical shift difference for a standard solution of ATP and its Mg complex at pH 7.0 was 10.99 ± 0.01 and 8.43 ± 0.01. In perfused beating guinea-pig hearts with an intracellular pH of 7.0 the difference was 8.43 ± 0.1, thus indicating a free Mg^{2+} concentration of 2.5 ± 0.7 mM[194]. This is only one example of a large number of studies on many different cell types[153] and it illustrates the ease with which the intracellular Mg^{2+} concentrations can be determined.

9 ¹H NMR OF METABOLITES

9.1 General

¹H NMR spectroscopy offers two important features for studying biological samples, namely the great inherent relative receptivity of the hydrogen nucleus and its ubiquity. Most of the applications of ¹H NMR have been to erythrocytes[10,43,195], and many of the procedures developed to study them can be applied, in principle, to other cells and fluids.

9.2 Sample preparation

Erythrocytes are amongst the most robust of mammalian cells so they are readily prepared for NMR spectroscopy by centrifugal washing; 2000 g for 10 min is adequate (e.g. Ref. 50). If the isotopic composition of the solvent is not important for the analysis then ^1H_2O signal suppression is aided by using buffers consituted in ^2H_2O (Section 2.2.1). Water exchange across the red cell membrane is fast[10] and ca 90 % exchange with ^2H_2O is achieved routinely with three washes with ca 5 volumes of buffer.

Sedimentation of the cells during the NMR experiment can lead to so-called 'settling artefacts' in measurements such as that of the intracellular water volume (Section 3), or any quantitative estimate that depends on peak areas rather than chemical shifts. Cell sedimentation can be avoided by using haematocrits (cytocrits) greater than ca 70 %[24] or by increasing the solvent density with impermeant arabinogalactan[196]. Some experiments require the use of low haematocrits, e.g. our measurements of the erythrocyte membrane permeability of DMMP[197]. We have observed that spinning the samples increases the rate of sedimentation of dilute suspensions.

If prolonged periods (ca 30 min) of NMR sampling are required on dilute cell suspensions, reproducible spectral intensities can be obtained by ensuring that the sample tube is filled, and that the base of the tube is well below the level of the receiver coil. A steady-state downflow of cells occurs and the cells pile up at the base of the tube; however, since the sedimentation is slow on the NMR time scale, the time-average cell density within the NMR coil is constant for a considerable time.

The quality of ^{1}H NMR spectra of erythrocytes depends on several parameters, an important one of which is the spin state of the iron in haemoglobin. Oxy- and carbomonoxyhaemoglobin are diamagnetic whereas deoxy- and methaemoglobin are paramagnetic[112,198]. Therefore, paramagnetic broadening of NMR signals can largely be avoided by saturating buffers and the erythrocyte samples with O_2 or CO; in fact, unless there is an experimental reason for not using it, we routinely gas samples with CO. Red cells are a special class of cell, using only anaerobic metabolism, so although CO is harmless to their metabolism bubbling with CO may be lethal for many other cell types. In red cells, the much higher affinity of haemoglobin for CO than O_2[199] ensures a sample that is diamagnetically stable for many hours. The CO also has the advantage of inhibiting the growth of many bacteria.

A final practical point is the need to pass all media through a Millipore filter; this not only sterilizes the medium but also removes particulate matter which may be paramagnetic. Phosphate buffers, in our experience, are particularly prone to be contaminated with 'paramagnetics' which are largely cleared by filtration. Alternatively, 0.1 mM EDTA usually relieves paramagnetic NMR peak broadening which may arise sporadically in some biological samples such as haemolysates.

9.3 The ^{1}H NMR resolution problem

9.3.1 General

Apart from the most serious NMR problem of eliminating the ^{1}H$_2$O resonance (Section 2), another is the necessity to observe resonances from small molecules in the face of much larger and broader peaks from non-exchanging protein protons. Suppression of protein ^{1}H resonances can be achieved in basically two ways. Firstly, saturation can be transferred through the protein spin system by cross-relaxation, giving rise to spin diffusion[200]. Secondly, the differences in T_2 values of the protons of large and small molecules can be exploited for spectroscopic discrimination.

9.3.2 Saturation transfer in haemoglobin

Partial suppression of the ^{1}H NMR haemoglobin envelope of erythrocytes may be achieved by applying a saturating decoupler r.f. field for ca 1 s at 7.1 ppm[201] or 6.9 ppm[43]; this irradiation is gated off and is then followed by a non-selective

^{1}H sampling pulse. The entire haemoglobin envelope is suppressed as a result of a negative nuclear Overhauser effect (NOE). The negative NOE arises when $\omega\tau_c > 1.12$, where ω is the Larmor frequency of the spins and τ_c is the dipolar correlation time[200]. In the process of spin diffusion the saturation is transferred by a series of mutual spin flips to other protons that are linked by dipolar interactions to the saturated protons[200]. If $\omega/2\pi = 400$ MHz a negative NOE results with molecules that have $\tau_c > 5 \times 10^{-10}$ s, and since typical values for proteins are 10^{-7}–10^{-9} s[203] spin diffusion is readily manifest in cells at high magnetic fields.

The saturating r.f. field is most effective at suppressing the protein signal envelope if it is applied at a frequency in the aliphatic region of the ^{1}H NMR spectrum; however, this spectral region is generally more crowded with peaks than the aromatic region so the latter is usually used. Direct irradiation of the ^{1}H$_2$O spins also leads to partial saturation of protein spins since some protein resonances are isochronous with ^{1}H$_2$O. Conversely, partial suppression of the ^{1}H$_2$O resonance automatically occurs during protein-proton saturation; this occurs because of spin exchange between the protein and the water which is closely associated with it inside cells. Therefore, if a metabolite has protons which rapidly exchange with those of water then the ^{1}H NMR signals of the metabolite may be suppressed by the saturation transfer process. For example, no resonances of DPG, the most abundant of the small molecules (other than water) in human erythrocytes, are observed, and the glutathione glycyl resonance is suppressed in spectra of samples constituted in ^{1}H$_2$O (Section 9.4.3)[24].

9.3.3   Spin–echo ^{1}H NMR

The application of the Hahn spin–echo pulse sequence[49] ($\pi/2$–τ–π–τ–acquire) was the key to applying ^{1}H NMR to studying cellular metabolism[24,204]. The use of the spin–echo method antedated that described above (Section 2.1.1) and is still the most general method for studying metabolism in erythrocytes and other tissues. The pulse sequence generates a signal echo of intensity $I(2\tau)$ at time 2τ after the $\pi/2$ pulse, and for a species undergoing isotropic unbounded diffusion in a linear magnetic field gradient (G) the signal intensity is described by

$$I(2\tau) = I(0)\exp\left(-\frac{2\tau}{T_2} - \frac{2\gamma^2 G^2 D\tau^3}{3}\right)F(J) \qquad (28)$$

where $I(0)$ is the signal intensity at $\tau = 0$, D is the translational diffusion coefficient and $F(J)$ is a term that describes the modulation of the signal due to homonuclear spin–spin coupling. The second term in equation 27 accounts for incomplete refocusing of the signal if the magnetic field (B_0) is inhomogeneous; this term can be ignored if D is effectively zero or if τ is small relative to the

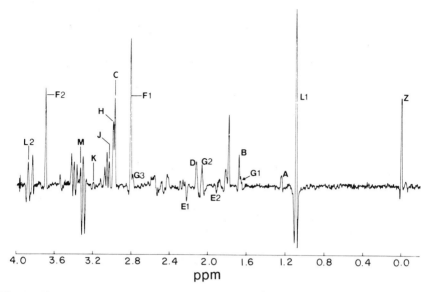

Fig. 14. The 4.0 to -0.2 ppm region of the 400 MHz ^{1}H NMR IRSE spectrum of rabbit brain homogenate; 512 transients were averaged into 16 384 memory locations. Resolution was enhanced by Gaussian multiplication. Chemical shifts in ppm referenced to TMS are given in parentheses. Peak assignments: A, alanyl methyl (1.240); B, acetate methyl (1.672); C, choline N-methyls (2.961); D, glutamate H$^\gamma$ (2.114); E1, glutamine H$^\gamma$ (2.218); E2, glutamine H$^\beta$ (1.908); F1, creatine methyl (2.798); F2, creatine methylene (3.687); G1, GABA H$^\beta$ (1.652); G2, GABA H$^\gamma$ (2.056); G3, GABA H$^\gamma$ (2.779); H, carnitine N-methyls (2.976); J, ergothioneine N-methyls (3.019); K, taurine H$^\alpha$ (3.189); L1, lactate methyl (1.089); L2, lactate H$^\alpha$ (3.871); M, glycine H^2 (3.322); and Z, TMS reference (0.000). From Ref. 207, with permission

translational correlation time τ_c ($\approx$ root mean square free displacement/$6D$) of the molecules in the sample.

From equation 28 it can be seen that the pulse sequence acts as a selective filter that allows the detection of signals from highly mobile solutes, while rejecting signals from molecules with long correlation times such as proteins. Spin–echo spectra of adrenal glands [205], erythrocytes[10,24,43], salivary glands[206], brain[207,208], muscle[209–212], tumour cells[213] and kidney[214] show resonances assigned to a multitude of metabolites. Figure 14, for example, is an IRSE (Section 2.3.2) spectrum obtained from rabbit brain homogenate; the spectral detail is great, with signals of some of the putative neurotransmitters being evident.

9.3.4 Phase modulation of signals in spin–echo spectra

In a spin–echo experiment the extent to which the signals from ^{1}H$_2$O and protein protons are suppressed, relative to those of small molecules, depends on the value of τ (Section 2.1.2). However, the value of τ determines the relative phases of components of multiplets. An excellent account of the basis of the phase modulation of multiplets has already been given by Rabenstein[43,215,216] so only a summary, without a physical explanation of the phenomena, will be given here. (a) A *singlet*, and the central component of a *triplet*, are always upright in the 'correctly' phased spectrum. (b) The two outer components (one each side) of a multiplet of order k ($\geqslant 2$) will be:

(i) *upright* if

$$\tau = \frac{i}{(k-1)J}, \quad \text{for } i = 1, \ldots, n \tag{29}$$

and (ii) *inverted* if

$$\tau = \frac{2i-1}{2(k-1)J}, \quad \text{for } i = 1, \ldots, n \tag{30}$$

where J is the coupling constant. (c) The inner components of a quartet obey the expression for a doublet. (d) These results apply to higher order multiplets where the separation of the components is a constant value, e.g. in a sextet the *central* three peaks can be viewed as a triplet, the outer components of which have phase behaviour in accordance with equations 29 and 30 with $k = 2$ and J being the separation (Hz) between the components.

The phase relationships of peaks in spin–echo spectra are a valuable assignment aid[10]. Decoupling of one spin from another, either by selective irradiation or by solvent exchange, changes the spin–echo modulation behaviour. For example, the lactate CH_3 resonance in a ^1H NMR spin–echo spectrum, obtained with $\tau = 60$ ms, of erythrocytes washed in ^2H_2O medium is upright[24] whereas for cells washed in ^1H_2O it is inverted[43]; the latter phenomenon is a result of the lactate CH_3 being coupled to its, $^1H^2$ but if the cells are in a medium containing ^2H_2O this atom exchanges with ^2H from the solvent, via glycolytic enzyme reactions, thus decoupling the CH_3 protons[217].

9.4 Quantification of ^1H NMR signals

9.4.1 General

There have been numerous studies of metabolizing systems in which it has been necessary to infer accurately concentrations from ^1H NMR signal amplitudes or intensities. The matter, statistically, is non-trivial and has been dealt with more extensively elsewhere but in slightly different contexts (see, e.g., Refs 218 and 219). An example in quantitative analytical clinical biochemistry, using ^{1}H NMR, is our own work on the assay of choline in the erythrocytes from manic depressive patients receiving the standard therapy for this condition, Li_2CO_3[220–222].

9.4.2   Choline assay in erythrocytes

The choline concentration was measured in haemolysates prepared from red cells that had been washed three times in ice-cold physiological saline. The samples were stored at $-12\,°C$ and upon thawing for NMR analysis the cells lysed spontaneously. An IRSE pulse sequence (Section 2.3.2) with $\tau_1 = \tau_2 = 60$ ms was used. In this experiment spectral peak amplitudes (height) are proportional to the concentration of the metabolite giving rise to it, and the amplitude is also a function of T_1, T_2 and J values[223]. In practice, the relationship between peak intensity (area) or amplitude and concentration is easily calibrated without the need to account for each of the above NMR parameters. For choline the detection limit, in a 400 MHz spectrometer, is about 0.01 mM.

The concentration of choline in haemolysates, or intact red cells, was determined by using a calibration graph that was constructed by adding known amounts (mass) of choline to haemolysates followed by measuring the corresponding peak intensities or amplitudes. In the case of intact cells, the peak amplitude of any metabolite including choline is also sensitive to the magnitude of magnetic field gradients; these arise as a result of differences in magnetic susceptibility between the intra- and extracellular compartments[112,132] (Section 9.3.2). The field gradient term of equation 28 often introduces unnecessary practical complications so it is useful to lyse the cells prior to analysis. Gassing the cells with CO obviates the field gradient problem with erythrocytes but will not be of general value for other cell types (Section 9.2). If enzyme assays are not required to be performed on the sample, then a perchloric acid extract yields a mixture which gives high-resolution ^{1}H NMR spectra devoid of a protein-proton envelope and the attendant problems of spectral assignment[211,214].

In the process of defining precise experimental and spectroscopic conditions for quantitative assays of choline[223] we found it necessary (a) to ensure that the receiver gain was carefully and reproducibly adjusted and (b) to ensure that no analogue-to-digital converter overload took place; if it did occur 'clipped' signals and baseline roll appeared, making precise peak amplitude estimates impossible. (c) The receiver r.f. filter-width also affected quantification[224], but the appropriate setting based on the spectral width is usually made automatically in modern spectrometers. (d) Samples had to be diamagnetically stable and therefore were gassed with either CO or O_2 (Section 9.2); the uniformity of sample pretreatment allowed the use of a standard calibration graph of choline $N^+(CH_3)_3$ peak amplitude versus concentration[223]. Additionally, (e) the standard deviations of the means of the estimates of choline peak areas were greater than those obtained by manually drawing baselines on the spectrum and then measuring peak amplitudes[223]. This phenomenon is, however, likely to depend on the proximity of other resonances to the one of interest and will not be a general finding. Indeed, because of the possibility that T_2^* may change during the period in which NMR measurements are made, it is theoretically preferable to measure peak areas.

The choline concentrations in erythrocytes were measured both by ^{1}H NMR spectroscopy and a modified assay based on choline oxidase; the assay values from both methods were always the same, within experimental error[223]. The choline levels were stable in samples stored at $-12\,°C$ for prolonged periods, provided the cells were originally washed in saline to remove Ca^{2+}, which activates endogenous phospholipase D. This enzyme catalyses the hydrolysis of lecithin, thus releasing choline from the cell membranes in the haemo-lysates[225]. The question of why our estimates of red cell choline are higher than those reported by other workers (albeit measured by less direct methods) is not yet fully resolved[226,227].

A final complication with quantitative ^{1}H NMR is 'biological' rather than instrumental; a non-uniform distribution of a solute can exist in a cell population. A specific example is human erythrocytes, which increase in density as they 'mature' during their lifetime of ca 120 days. The choline concentration in density-fractionated cells varies uniformly with cell age[223]. Also, the creatine levels decline exponentially with cell age and result in a dramatic 10-fold difference in concentration between the youngest and oldest cells[228].

9.4.3 Peptide concentrations

^{1}H NMR spectra of glutathione (GSH) in media of differing pH are shown in Figure 15. In ^{1}H$_2$O at physiological pH values the exchange rate of the reaction between the amide protons of GSH and water is intermediate to slow (at $\mathbf{B}_0 = 9.4$ T); the glycyl methylene protons (H^2) are coupled ($J_{H^N H^\alpha}$) to the amide proton (H^N), thus splitting the resonance into a doublet. In a spin–echo experiment this doublet is phase modulated, as is evident in Figure 16 where the ^{1}H spectra of haemolysates prepared in ^{1}H$_2$O differ from those of haemolysates prepared in ^{2}H$_2$O. In this type of experiment, if the sample is irradiated at the ^{1}H$_2$O resonance frequency then saturation transfer takes place to the glycyl H^2 protons, thus partially suppressing the resonance (Section 9.3.2). Further, since this saturation transfer may not be reproducibly applied from one sample to the next it is impossible, by using this methodology, to use the glycyl H^2 to monitor quantitatively a metabolic reaction involving GSH in ^{1}H$_2$O.

The effect of solvent on enzyme catalytic rate, a so-called solvent kinetic isotope effect (KIE), can be used to obtain an insight into the molecular mechanism of the enzyme[29]. Accordingly, the time course of hydrolysis of

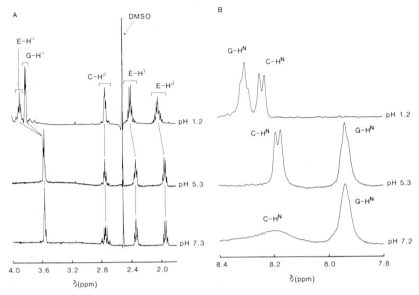

Fig. 15. ^{1}H NMR (400 MHz) spectra of glutathione (GSH) in ^1H_2O. (A) Aliphatic region of the ^{1}H NMR spectrum obtained with the peptide in ^1H_2O, at 37 °C for pH 5.3 and 7.2 and at 25 °C for pH 1.2. The spectra (64 transients) were obtained using gated irradiation of the ^1H_2O resonance. (B) Backbone amide region of the ^{1}H NMR spectrum of GSH obtained from the above solutions. The spectrum from the solution at pH 1.2 was obtained from 64 transients with gated r.f. irradiation at the ^1H_2O resonance frequency; those spectra obtained at higher pH were from 16 transients with the Redfield and Gupta[58] '214' pulse sequence [Section 2.4.2]. Assignments: E, glutamyl; G, glycyl; and C, cysteinyl proton. From Ref. 31, with permission

γ-glutamylalanine by a haemolysate, which contains the endogenous enzyme γ-glutamyl amino acid cyclotransferase, was studied and was found to be different in ^1H_2O and ^2H_2O media[27]. This kinetic information was then used in formulating a model of the mechanism of the isolytic process in the active site of γ-glutamyl amino acid cyclotransferase[24].

Figure 17 shows a typical series of ^{1}H NMR spectra obtained from a haemolysate using the spin-echo pulse sequence ($\tau = 60$ ms) with homogated irradiation (Section 2) at the ^1H_2O resonance frequency[229]. The peptide L-glutamyl-L-alanine (Glu-Ala) was also present in the haemolysate, and it was hydrolysed by endogenous peptidases; the hydrolysis is evident from the declining amplitude of the glutamyl H$^\gamma$ and alanyl H$^\beta$ resonances of the peptide, and concomitant appearance of free glutamate H$^\gamma$ (b) and alanine H$^\beta$ (d) resonances. Thus progress curves of substrate and product concentration ($[S]_t$ and $[P]_t$), can be obtained from such an experiment. However, it is imperative to determine a scaling relationship between peak amplitude and solute concentra-

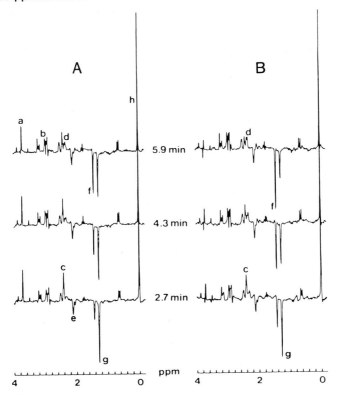

Fig. 16. Cleavage of γ-glutamylalanine (γ-Glu-Ala) by human erythrocyte lysates in ^1H_2O and ^2H_2O; the effect of $J_{H^NH^2}$-dependent modulation of the ^1H spin–echo signal of glutathione (GSH) glycyl H^2 (peak a). IRSE (Section 2.3.2) spectra (400 MHz; 32 transients) of an undiluted erythrocyte lysate at 37°C with haematocrit before lysis of 0.83. (B) The cells were washed in ^1H_2O Krebs buffer; (A) the final wash was in a ^2H_2O Krebs buffer yielding a final $^1H_2O:^2H_2O$ of 1:1. At time 0 min, 7.5 μl of γ-Glu-Ala (0.28 M) were added to the sample (0.5 ml). Peak assignments: a, GSH glycyl H^2; b, creatine methyl; c, γ-Glu-Ala glutamyl $H^γ$; d, 5-oxoproline $H^γ$; e, GSH and γ-Glu-Ala glutamyl $H^β$; f, alanine $H^β$; g, γ-Glu-Ala alanyl $H^β$; h, methyl of the external reference 2,2-dimethyl-2-silapentane-5-sulphonate. From Ref. 27, with permission

tion. The scaling factors can be obtained by using the principle of 'conservation of mass':

$$[S]_t + [P]_t = [S]_0 \qquad (31)$$

where $[S]_0$ is the initial peptide concentration. Equation 31 is expressed in the form[229]

$$\varepsilon_s S + \varepsilon_p P = [S]_0 \qquad (32)$$

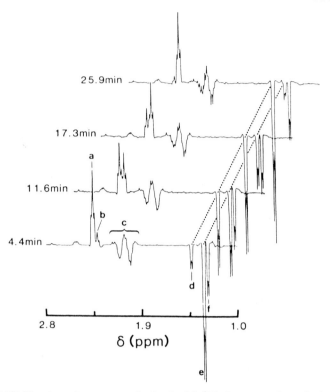

Fig. 17. ^{1}H NMR spin–echo spectra obtained with ^{1}H$_2$O suppression of a haemolysate (original haematocrit = 0.73) to which Glu-Ala was added ([S]$_0$ = 13.7 mM) at zero time. Spectral assignments; a, glutamyl H$^\gamma$; b, glutamate H$^\gamma$; c, glutamate/glutamyl H$^\beta$; d, alanine H$^\beta$; alanyl H$^\beta$; f, lactate methyl. From Ref. 230, with permission

where S and P are the spectral peak amplitudes of substrate and product, respectively, and ε_s and ε_p are molar absorption coefficients. Rearranging equation 32 gives

$$S = (-\varepsilon_p/\varepsilon_s)P + [S]_0/\varepsilon_s \tag{33}$$

Regression of a line, $y = mx + b$, on to data pairs of substrate versus product-peak amplitude yields the 'best fit' NMR absorption coefficients and thus allows the calculation of the solute concentrations.

This method of determining ε_p and ε_s values is superior to that involving a calibration line since the latter assumes a unique value of coefficients for each solute, whereas these may vary between samples if different paramagnetic states of the sample exist, or local magnetic field gradients differ between samples.

By using this approach the degradation pathway of the pentapeptide enkephalin by human erythrocyte lysates was determined[230], as were the kinetic characteristics of porcine kidney prolidase[231] and, in less detail, some other red cell enzymes[26].

9.5 ¹H NMR of body fluids

9.5.1 Urine

A vast amount of biochemical information is available on the clinical state of a patient from the analysis of urine; urine is also the most readily available body fluid. It is, in principle, feasible to monitor ^{19}F-labelled drug excretion by ^{19}F NMR of urine. There are normally no ^{39}P-containing metabolites in urine, and ^{13}C NMR spectra must usually rely on the natural abundance ^{13}C in various compounds, hence the spectra takes too long to acquire for routine clinical biochemical use. Therefore, NMR studies of urine are primarily restricted to the ^{1}H nucleus. The paramount problem with the ^{1}H NMR of urine is the water signal; however, this signal can be adequately suppressed by using the procedures described in Section 2. The next task that besets the clinical biochemist who is confronted with a ^{1}H NMR spectrum of urine is the spectral assignment of the plethora of excretion products. This task is fortunately aided by the comprehensive background information made available through the work of Sadler, Nicholson and collaborators. The spectral assignments have been carried out by using information on spin–echo peak modulation patterns (Section 9.3.4), peak multiplicities, decoupling experiments and in some cases 2D COSY connectivities[232].

^{1}H NMR of urine has been used to contribute understanding to the diagnosis of various clinical states, including diabetes[233] and fasting[234]. It has been used to monitor metabolic disorders such as organic acidurias[235,236], biotinidase deficiency, glutaricaciduria, hyperglycinaemia type II, isovaleric acidaemia, congenital lactic acidosis, 3-methylcrotonyl-CoA carboxylase defect, β-keto-thiolase deficiency, methylmalonic aciduria, phenylketonuria and maple syrup urine disease[237], and, independently, phenylketonuria[238] and maple syrup urine disease[239]. Metabolism of the drug acetaminophen has been extensively studied using ^{1}H NMR of urine to monitor metabolic products[232]. The metabolic products of the antiprotozoal–antiseptic agent metranidazole can also be detected in urine by ^{1}H NMR[240]. Finally, the ^{1}H NMR procedure has been used to follow the effects of renal damage on urinary composition[8,241,242].

9.5.2 Plasma and serum

High-resolution ^{1}H NMR spin–echo spectra of human plasma can be readily obtained within a few minutes and the spectra display a vast amount of detail; specifically, prominent signals are evident from glucose, a range of amino acids, lactate, fatty acids and carbohydrate moieties of glycoproteins[243-246]. Further, by measuring the chemical shifts of EDTA resonances it is possible to infer the plasma Ca^{2+} and Mg^{2+} concentrations[244]. At $B_0 = 9.4$ T, the Ca^{2+} and Mg^{2+} complexes of EDTA are in slow exchange and characteristic spectral features are observed for the ethylenic protons. Thus, by using a method of standard additions to establish a calibration relationship, quantification of the ions in plasma can be made.

A particular clinical example which emphasizes the great value of the ^1H NMR spectra of plasma, for identifying unsuspected compounds which may accummulate in disease states, is as follows. A patient recovering from jejuno-ileal bypass surgery developed an organic acidosis. Conventional assays for L-lactate did not detect elevated levels of this compound. A ^1H NMR spectrum of the patient's plasma, however, revealed a very prominent lactate CH_3 resonance. Quantitative ^1H NMR analysis for total lactate revealed D-lactate to be the source of the acidosis because the total lactate level was well above that determined by conventional stereospecific enzyme assay[247].

9.5.3   Cerebrospinal fluid

The ^1H NMR procedures applied to urine and plasma samples are also relevant to cerebrospinal fluid (CSF). The method requires a 0.5 ml of sample, and this volume is well within the limits of fluid normally obtained at lumbar puncture. As was the case with plasma and urine, 1D and 2D NMR spectroscopy have been used to characterize and quantify the spectra of a large number of metabolites in CSF[248,249].

9.6 ^1H NMR in tumour screening

The ^1H_2O-suppressed spin-echo spectrum of plasma has prominent resonances that are assigned to plasma lipoprotein lipids. Fossel et al.[250] measured the linewidths of the two composite peaks, centred at ca 0.77 and ca 1.9 ppm, under rigorously controlled experimental and data-processing conditions. They found systematic variations between the line width ($\Delta v_{1/2}$) in the plasma samples of patients suffering from various diseases and in various clinical states. Of most interest was the observation that the mean $\Delta v_{1/2}$ values measured in spectra of plasma from patients with malignant tumours (29.9 ± 2.52 Hz, $n = 81$) was significantly different from those in normal controls (39.5 ± 1.6 Hz, $n = 44$). There was, however, some 'overlap' of values with other groups, specifically pregnant women and patients with previously treated malignancies and a class of benign tumour. The simple ^1H NMR-based test may nevertheless form the basis of a procedure for detecting preclinical cancer; however, much more work is required in order to establish the frequency of false-positives and false-negatives in the test.

The observation of altered lipoprotein-^{1}H relaxation rates may be related to conclusions drawn in other work which suggests that cell changes in membrane lipid mobility accompany the development of the metastatic state in the progressive development of malignant tumours[251,252]. These mobility changes have been inferred from ^{1}H NMR relaxation-time measurements made on the overlapping resonances of lipids in membranes isolated from the tumour cells. The question of which membranes give rise to these ^{1}H NMR lipid signals in the spectra and whether it is, in fact, really the plasma membranes, has begun to be addressed[253]. Further applications of this quantitative analysis are required before a definite claim as to the source of the signals can be made.

10 ^{31}P, ^{13}C AND OTHER NUCLEI

The NMR-receptive nucleus which has been most used as a probe of biochemical reactions in cells is ^{31}P. The general principles of sample preparation, discussed in Section 9 for ^{1}H NMR, also apply to ^{31}P and other nuclei. Obviously there is no ^{1}H$_2$O resonance problem with spectra of other nuclei except in the case of some 'reverse' polarization transfer experiments (Section 2.2.4). The vast bulk of papers pertaining to ^{31}P NMR of cells have as an important component either the measurement of pH (Section 4.2), or the measurement of the 'energy status' of the cells. An indicator of the latter can be obtained from the ratio of [ATP] to [ADP] and is derived from the expression

$$[ATP]/[ADP] = \gamma/(\gamma - \beta) \tag{34}$$

where β and γ are, respectively, the signal intensities of the βP of ATP (ca 20 ppm) and the unresolved combined resonances of the γP of ATP and βP of ADP[12]. The experimental requirements for obtaining reproducible quantitative ^{31}P spectra of cells and tissues have been dealt with in several places elsewhere (e.g. Ref. 12) so they will not be discussed further here; the factors are all closely related to those discussed in Section 9. One important aspect is the avoidance of progressive saturation of the spins by using small flip angles when accumulating many transients for the purpose of signal averaging. This is also particularly important when attempting to quantify metabolites, or in simply making assignments, of ^{13}C spectra, since the carboxyl and carbonyl ^{13}C nuclei have particularly long T_1 values. These values can be at least 40 s, thus implying the need for very slow recycling in spectral acquisition. Also, slow recycling is particularly important in enzyme and transport kinetic studies where rate constants of exchange processes are evaluated on the basis of peak areas (e.g. Ref. 219).

Although limits of space and 'emphasis' have allowed only scant attention to be given here to ^{13}C NMR studies of metabolism, it should be stated that the technique has been used to monitor biochemical reactions in many cell types, in yeast[254,255] and heart[256] using ^{13}C-labelled substrates and by monitoring natural abundance ^{13}C in liver glycogen[257] and fat in adipose tissue[258]. Both glycogen in liver and fat in adipose tissue represent large pools of carbon nuclei resonating in a small band of frequencies; nevertheless, changes in diet and dietary state can in principle be detected *in vivo* within a few minutes using only the signals from natural abundance ^{13}C in glycogen and fat. This result is promising for studies of aspects of lipid and glycogen metabolism *in vivo* by using whole-human spectroscopy.

11   CONCLUSION

In the 15 years since the first report of the use of NMR to follow metabolic processes there has been a divergence of applications from single cells to more complex cells, organs, tissues and whole animals; there has also been a divergence in the use of NMR-receptive nuclei for probing various aspects of metabolism.

The magnetic field strengths currently available for NMR spectrometers render the technique insensitive to the many small molecules and some of the ions important in metabolic regulation. Nevertheless, the success of the ^{31}P NMR indirect detection of H^+ elicits some hope for the indirect detection of that most important of the intracellular metal ions, Ca^{2+}.

As an analytical tool for the non-specific identification of metabolites and drugs in blood cells, plasma, urine and CSF, ^1H NMR is almost unparalleled and, except for its low sensitivity (relative to HPLC, say) and high initital capital cost of the instrument, many more spectrometers might have already been installed in clinical biochemistry departments.

The many NMR methods surveyed here for obtaining spectra from biological samples represent the beginning of one of the most rapidly evolving technical facets of modern biochemistry.

REFERENCES

1. E. Odeblad, B. N. Bhar and G. Lindstrom, *Arch. Biochem. Biophys.*, **63**, 221 (1956).
2. O. Jardetzky and J. E. Wertz, *Am. J. Physiol.*, **187**, 608 (1956).
3. R. Mathur-De Vré, *Prog. Biophys. Mol. Biol.*, **35**, 103 (1979).
4. R. N. Moon and J. H. Richards, *J. Biol. Chem.*, **248**, 7276 (1973).
5. D. I. Hoult, S. J. W. Busby, D. G. Gadian, G. K. Radda, R. E. Richards and P. J. Seeley, *Nature* (*London*), **252**, 285 (1974).

6. T. O. Henderson, A. J. R. Costello and A. Omachi, *Proc. Natl. Acad. Sci. USA*, **71**, 2487 (1974).
7. C. T. Burt, T. Glonek and M. Bárány, *J. Biol. Chem.*, **251**, 2584 (1976).
8. D. G. Gadian, G. K. Radda, R. E. Richards and P. J. Seeley. In R. G. Shulman (ed.), *Biological Applications of Magnetic Resonance* Academic Press, New York, 1979 p. 463.
9. K. Ugurbil, R. G. Shulman and T. R. Brown. In R. G. Shulman (ed.) *Biological Applications of Magnetic Resonance*, Academic Press, New York, 1979 p. 537.
10. P. W. Kuchel, *CRC. Crit. Rev. Anal. Chem.*, **12**, 155 (1981).
11. D. G. Gadian and G. K. Radda, *Annu. Rev. Biochem.*, **50**, 69 (1981).
12. D. G. Gadian, *Nuclear Magnetic Resonance and its Applications to Living Systems*, Clarendon Press, Oxford, 1982.
13. R. G. Shulman, *Sci. Am.*, **248**, 86 (1983).
14. J. K. M. Roberts, *Annu. Rev. Plant Physiol.*, **35**, 375 (1984).
15. M. J. Avison, H. P. Hetherington and R. G. Shulman, *Annu. Rev. Biophys. Biophys. Chem.*, **15**, 377 (1986).
16. R. S. Norton, M. A. MacKay and L. J. Borowitzka, *Biochem. J.*, **202**, 699 (1982).
17. Z. H. Endre, B. E. Chapman and P. W. Kuchel, *Biochem. J.*, **216**, 655 (1983).
18. Z. H. Endre and P. W. Kuchel, *Biophys. Chem.*, **24**, 337 (1986).
19. R. S. Norton, *Bull. Magn. Reson.*, **3**, 29 (1980).
20. A. I. Scott and R. L. Baxter, *Annu. Rev. Biophys. Bioeng.*, **10**, 151 (1981).
21. A. G. Redfield. In E. M. Bradbury and C. Nicolini (eds) *NMR in the Life Sciences*, Plenum Press, New York, 1985, p. 1.
22. A. Abragam, *The Principles of Nuclear Magnetism*, Clarendon Press, Oxford, 1978.
23. J. Sandström, *Dynamic NMR Spectroscopy*, Academic Press, London, 1982.
24. F. F. Brown, I. D. Campbell, P. W. Kuchel and D. L. Rabenstein, *FEBS Lett.*, **82**, 12 (1977).
25. S. J. Hyslop, G. F. King and P. W. Kuchel, *Am. J. Hematol.*, **16**, 183 (1987).
26. P. W. Kuchel, B. E. Chapman, Z. H. Endre, G. F. King, D. R. Thorburn and M. J. York, *Biomed. Biochim. Acta*, **43**, 719 (1984).
27. M. J. York, P. W. Kuchel and B. E. Chapman, *J. Biol. Chem.*, **259**, 15085 (1984).
28. D. R. Thorburn and P. W. Kuchel, *Eur. J. Biochem.*, **150**, 371 (1985).
29. W. W. Cleland, *CRC Crit. Rev. Biochem.*, **13**, 385 (1982).
30. K. Wüthrich, *NMR in Biological Research: Peptides and Proteins*, North Holland, Amsterdam, 1976.
31. M. J. York, G. R. Beilharz and P. W. Kuchel, *Int. J. Pept. Protein Res.*, **29**, 638 (1987).
32. D. L. Rabenstein, S. Fan and T. T. Nakashima, *J. Magn. Reson.*, **64**, 541 (1985).
33. S. Meiboom and D. Gill, *Rev. Sci. Instrum.*, **29**, 688 (1958).
34. D. L. Rabenstein, G. S. Srivatsa and R. W. K. Lee, *J. Magn. Reson.*, **71**, 175 (1987).
35. S. Connor, J. Everett and J. K. Nicholson, *Magn. Reson. Med.*, **4**, 461 (1987).
36. M. T. Emerson, E. Grunwald and R. A. Kromhout, *J. Chem. Phys.*, **33**, 547 (1960).
37. D. L. Rabenstein and A. A. Isab, *J. Magn. Reson.*, **36**, 281 (1979).
38. E. G. Finer, F. Franks and M. J. Tait, *J. Am. Chem. Soc.*, **94**, 4424 (1972).
39. R. L. Vold, E. S. Daniel and S. Chan, *J. Am. Chem. Soc.*, **92**, 6671 (1970).
40. A. G. Redfield, *Methods Enzymol.*, **49**, 253 (1978).
41. J. P. Jesson, P. Meakin and G. Kniessel, *J. Am. Chem. Soc.*, **95**, 618 (1973).
42. I. D. Campbell, C. M. Dobson, G. Jeminet and R. J. P. Williams, *FEBS Lett.*, **49**, 115 (1974).
43. D. L. Rabenstein, *J. Biochem. Biophys. Methods*, **9**, 277 (1984).
44. R. K. Gupta, *J. Magn. Reson.*, **24**, 461 (1976).
45. I. D. Campbell, C. M. Dobson and R. G. Ratcliffe, *J. Magn. Reson.*, **27**, 455 (1976).

46. S. L. Patt and B. D. Sykes, *J. Chem. Phys.*, **56**, 3182 (1972).

47. F. W. Benz, J. Feeney and G. C. K. Roberts, *J. Magn. Reson.*, **8**, 114 (1972).

48. R. L. Vold, J. S. Waugh, M. P. Klein and D. E. Phelps, *J. Chem. Phys.*, **48**, 3821 (1968).

49. E. L. Hahn, *Phys. Rev.*, **80**, 580 (1950).

50. M. J. York, P. W. Kuchel, B. E. Chapman and A. L. Jones, *Biochem. J.*, **207**, 65 (1983).

51. G. F. King and P. W. Kuchel, *Biochem. J.*, **220**, 553 (1984).

52. R. A. Dwek, *Nuclear Magnetic Resonance (N.M.R.) in Biochemistry, Applications to Enzyme Systems*, Clarendon Press, Oxford, 1973.

53. T. Ogino, Y. Arata, S. Fujiwara, H. Shaun and T. Beppu, *J. Magn. Reson.*, **31**, 523 (1978).

54. J. Dadok and R. F. Sprecher, *J. Magn. Reson.*, **13**, 243 (1974).

55. R. K. Gupta, J. A Ferretti and E. Becker, *J. Magn. Reson.*, **13**, 275 (1974).

56. P. W. Kuchel, *Anal. Biochem.*, **88**, 37 (1978).

57. J. D. Glickson, R. Rowan, T. P. Pitner, J. Dadok, A. A. Bothner-By and R. Walter, *Biochemistry*, **15**, 1111 (1976).

58. A. G. Redfield and R. K. Gupta, *Adv. Magn. Reson.*, **5**, 81 (1971).

59. A. G. Redfield. In *NMR: Basic Principles and Progress*, P. Diehl, E. Flück and R. Kosfeld (eds.) Springer-Verlag, Gottingen, **13**, 1976, p. 152.

60. G. M. Clore, B. J. Kimber and A. M. Gronenborn, *J. Magn. Reson.*, **54**, 170 (1983).

61. P. Plateau, C. Dumas and M. Gueron, *J. Magn. Reson.*, **54**, 46 (1983).

62. V. Sklenár and Z. Starcuk, *J. Magn. Reson.*, **50**, 495 (1982).

63. M. M. Pike, J. C. Frazer, D. F. Dedrick, J. S. Ingwall, P. D. Allen, C. S. Springer and T. W. Smith, *Biophys. J.*, **48**, 159 (1985).

64. P. J. Hore, *J. Magn. Reson.*, **54**, 539 (1983).

65. P. J. Hore, *J. Magn. Reson.*, **55**, 283 (1983).

66. C. Wang and A. Pardi, *J. Magn. Reson.*, **71**, 154 (1987).

67. M. R. Spiegel, *Schaum's Outline Series, Theory and Problems of Fourier Analysis*, McGraw-Hill, New York, 1974.

68. R. N. Bracewell, *The Fourier Transform and its Applications*, McGraw-Hill, London, 1978.

69. D. C. Champeney, *Fourier Transforms in Physics*, Adam Hilger, Bristol, 1985.

70. D. L. Turner, *J. Magn. Reson.*, **54**, 146 (1983).

71. G. A. Morris, K. I. Smith and J. C. Waterton, *J. Magn. Reson.*, **68**, 526 (1986).

72. M. H. Levitt and M. F. Roberts, *J. Magn. Reson.*, **71**, 576 (1987).

73. R. Freeman, T. H. Mareci and G. A. Morris, *J. Magn. Reson.*, **42**, 341 (1981).

74. M. R. Bendall, D. T. Pegg and D. M. Doddrell, *J. Magn. Reson.*, **45**, 8 (1981).

75. M. R. Bendall, D. T. Pegg, D. M. Doddrell and J. Field, *J. Magn. Reson.*, **51**, 520 (1983).

76. J. M. Bulsing, W. M. Brooks, J. Field and D. M. Doddrell, *Chem. Phys. Lett.*, **104**, 229 (1984).

77. W. M. Brooks, M. G. Irving, S. J. Simpson and D. M. Doddrell, *J. Magn. Reson.*, **56**, 521 (1984).

78. W. M. Brooks, *Heteronuclear Multipulse and Biological Applications of Nuclear Magnetic Resonance Spectroscopy*, PhD Thesis, Griffith University, Queensland, 1984.

79. C. L. Dumoulin and E. A. Williams, *J. Magn. Reson.*, **66**, 86 (1986).

80. D. P. Burrum and R. R. Ernst, *J. Magn. Reson.*, **39**, 163 (1980).

81. H. Rottenberg, *Methods Enzymol.*, **25**, 547 (1979).

82. N. Mohondas, Y. R. Kim, D. H. Tyko, J. Orlik, J. Wyatt and W. Groner, *Blood*, **68**, 506 (1986).

83. B. M. Rayson and R. K. Gupta, *J. Biol. Chem.*, **260**, 7276 (1985).
84. D. Hoffman and R. K. Gupta, *J. Magn. Reson.*, **70**, 481 (1986).
85. D. Hoffman, A. M. Kumar, A. Spitzer and R. K. Gupta, *Biochim. Biophys. Acta*, **889**, 355 (1986).
86. S. Ogawa, C. C. Boens and T. M. Lee, *Arch. Biochem. Biophys.*, **210**, 740 (1981).
87. H. Shinar and G. Navon, *Biophys. Chem.*, **20**, 275 (1984).
88. H. Shinar and G. Navon, *FEBS Lett.*, **193**, 75 (1985).
89. B. E. Cowan, D. Y. Sze, M. T. Mai and O. Jardetzky, *FEBS Lett.*, **184**, 130 (1985).
90. K. Kirk and P. W. Kuchel, *J. Magn. Reson.*, **62**, 568 (1985).
91. K. Kirk and P. W. Kuchel, *Stud. Biophys.*, **116**, 139 (1986).
92. J. E. Raftos, K. Kirk and P. W. Kuchel, *Biochim. Biophys. Acta*, **968**, 160 (1988).
93. A. J. Grimes, *Human Red Cell Metabolism*, Blackwell Scientific, Oxford, 1980.
94. J. Bernhardt and H. Pauly, *J. Phys. Chem.*, **79**, 584 (1975).
95. K Kirk, P. W. Kuchel and R. J. Labotka, *Biophys. J.*, **54**, 241 (1988).
96. R. J. Labotka and A. Omachi, *Biomed. Biochim. Acta*, **46**, 560 (1987).
97. R. J. Labotka and A. Omachi, *J. Biol. Chem.*, **262**, 305 (1987).
98. I. M. Stewart, B. E. Chapman, K. Kirk, P. W. Kuchel, V. A. Lovric and J. E. Raftos, *Biochim. Biophys. Acta*, **885**, 23 (1986).
99. K. Kirk and P. W. Kuchel, unpublished results.
100. K. Kirk, J. E. Raftos and P. W. Kuchel, *J. Magn. Reson.*, **70**, 484 (1986).
101. R. Iles, *Biosci. Rep.*, **1**, 687 (1981).
102. D. G. Gadian, *Biosci. Rep.*, **1**, 449 (1981).
103. J. K. M. Roberts and O. Jardetzky, *Biochim. Biophys. Acta*, **639**, 53 (1981).
104. D. G. Gadian, G. K. Radda, M. J. Dawson and D. R. Wilkie. In R. Nuccitelli and D. W. Deamer (eds), *Intracellular pH; its Measurement, Regulation and Utilization in Cellular Function*, Alan R. Liss, New York, 1982.
105. M. Bárány and T. Glonek. In D. G. Gorenstein (ed.), *Phosphorus-31 NMR, Principles and Applications*, Academic Press, Orlando, 1984, Ch. 17.
106. J. K. M. Roberts and O. Jardetzky, *Biochim. Biophys. Acta*, **639**, 53 (1981).
107. J. K. M. Roberts, N. Wade-Jardetzky and O. Jardetzky, *Biochemistry*, **20**, 5389 (1981).
108. J. L. Slonczewski, B. P. Rosen, J. R. Alger and R. M. Macnab, *Proc. Natl. Acad. Sci. USA*, **78**, 6271 (1981).
109. R. J. Labotka and R. A. Kleps, *Biochemistry*, **22**, 6089 (1983).
110. J. E. Raftos, B. E. Chapman, P. W. Kuchel, V. A. Lovric and I. M. Stewart, *Haematologia*, **19**, 251 (1986).
111. A. Peterson, J. P. Jacobsen and M. Horder, *Magn. Reson. Med.*, **4**, 341 (1987).
112. M. Fabry and R. C. San George, *Biochemistry*, **22**, 4119 (1983).
113. E. C. C. Lin, *Annu. Rev. Microbiol.*, **30**, 535 (1976).
114. W. J. Thoma, J. G. Steiert, R. L. Crawford and K. Ugurbil, *Biochem. Biophys. Res. Commun.*, **138**, 1106 (1986).
115. R. K. Deuel, G. M. Yue, W. R. Sherman, D. J. Schickner and J. J. H. Ackerman, *Science*, **228**, 1329 (1985).
116. J. A. Bailey, S. R. Williams, G. K. Radda and D. G. Gadian, *Biochem. J.*, **196**, 171 (1981).
117. W. M. Brooks and R. J. Willis, *J. Mol. Cell. Cardiol.*, **17**, 747 (1985).
118. J. K. M. Roberts, P. M. Ray, N. Wade-Jardetzky and O. Jardetzky, *Nature (London)*, **283**, 870 (1980).
119. R. J. Gillies, T. Ogino, R. G. Shulman and D. C. Ward, *J. Cell. Biol.*, **95**, 24 (1982).
120. J. B. Martin, R. Bligny, F. Rebeille, R. Dowe, J. J. Leguay, Y. Mathieu and J. Guern, *Plant Physiol.*, **70**, 1156 (1982).
121. M. J. Kushmerick and R. A. Meyer, *Am. J. Physiol.*, **248**, C542 (1985).

122. S. M. Cohen, S. Ogawa, H. Rottenberg, P. Glynn and T. Yamane, *Nature (London)*, **273**, 554 (1978).
123. H. B. Pollard, H. Shindo, C. E. Creutz, C. J. Pazoles and J. S. Cohen, *J. Biol. Chem.*, **254**, 1170 (1979).
124. L. H. Schliselfield, C. T. Burt and R. J. Labotka, *Biochemistry*, **21**, 317 (1982).
125. F. F. Brown and I. D. Campbell, *FEBS Lett.*, **65**, 322 (1976).
126. J. L. Nieto, *FEBS Lett.*, **136**, 85 (1981).
127. D. L. Rabenstein and A. A. Isab, *Anal. Biochem.*, **121**, 423 (1982).
128. K. Yoshizaki, Y. Seo and H. Nishikawa, *Biochim. Biophys. Acta*, **678**, 283 (1981).
129. C. Arús and M. Bárány, *Biochim. Biophys. Acta*, **886**, 411 (1986).
130. Y. Seo, K. Yoshizaki and T. Morimoto, *Jpn. J. Physiol.*, **33**, 721 (1983).
131. Y. Seo, *Am. J. Physiol.*, **297**, C175 (1984).
132. K. M. Brindle, F. F. Brown, I. D. Campbell, C. Grothwohl and P. W. Kuchel, *Biochem. J.*, **180**, 37 (1979).
133. J. Homer, *J. Magn. Reson.*, **57**, 171 (1984).
134. K. Ugurbil, M. H. Fukami and H. Holmsen, *Biochemsitry*, **23**, 416 (1984).
135. J. A. Cramer and J. H. Prestegard, *Biochem. Biophys. Res. Commun.*, **75**, 295 (1977).
136. M. A. Stidham, D. E. Moreland and J. N. Siedow, *Plant Physiol.* **73**, 517 (1983).
137. P. W. Kuchel, B. T. Bulliman, B. E. Chapman and K. Kirk, *J. Magn. Reson.*, **74**, 1 (1987).
138. J. S. Taylor, C. Deutsch, G. G. McDonald and D. F. Wilson, *Anal. Biochem.*, **114**, 415 (1981).
139. J. S. Taylor and C. Deutsch, *Biophys. J.*, **43**, 261 (1983).
140. C. Deutsch, J. S. Taylor and D. F. Wilson, *Proc. Natl. Acad. Sci. USA*, **79**, 7944 (1982).
141. T. Kashiwayura, C. Deutsch, J. Taylor, M. Erecinska and D. F. Wilson, *J. Biol. Chem.*, **259**, 237 (1984).
142. M. M. Civan, L. E. Lin, K. Peterson-Yantorno, J. Taylor and C. Deutsch, *Am. J. Physiol.*, **247**, C506 (1984).
143. J. C. Metcalfe, T. R. Hesketh and G. A. Smith, *Cell Calcium*, **6**, 183 (1985).
144. J. Hoffman, *Circulation*, **26**, 1201 (1962).
145. G. E. Lindenmayer and A. Schwartz, *J. Mol. Cell. Cardiol.*, **7**, 591 (1975).
146. P. K. Lauf, *Am. J. Physiol.*, **249**, 271 (1985).
147. M. D. Rhoda, F. Giraud, C. T. Craescu and Y. Beuzard, *Cell Calcium*, **6**, 397 (1985).
148. R. Y. Tsien, *Biochemistry*, **19**, 2396 (1980).
149. R. Y. Tsien, T. Pozzan and T. J. Rink, *Nature (London)*, **295**, 68 (1982).
150. R. T. Tsien, T. Pozzan and T. J. Rink, *J. Cell. Biol.*, **94**, 324 (1982).
151. R. K. Harris and B. E. Mann, *NMR and the Periodic Table*, Academic Press, London, 1978.
152. G. A. Smith, R. T. Hesketh, J. C. Metcalfe, J. Feeney and P. G. Morris, *Proc. Natl. Acad. Sci. USA*, **8**, 7178 (1983).
153. R. K. Gupta, P. Gupta and R. D. Moore, *Annu. Rev. Biophys. Bioeng.*, **13**, 221 (1984).
154. T. R. Hesketh, G. A. Smith, J. P. Moore, M. V. Taylor and J. C. Metcalfe, *J. Biol. Chem.*, **258**, 4876 (1983).
155. E. Murphy, L. Levy, L. R. Berkowitz, E. P. Orringer, S. A. Gabel and R. E. London, *Am. J. Physiol.*, **251**, C496 (1986).
156. L. A. Levy, M. E. Murphy and R. E. London, *Am. J. Physiol.*, **252**, C441 (1987).
157. B. Alberts, D. Bray, J. Lewis, M. Raff, K. Roberts and J. D. Watson, *Molecular Biology of the Cell*, Garland, New York, 1983.
158. J. M. Hamyln, R. Ringel, J. Schaeffer, P. D. Levison, B. P. Hamilton, A. A. Kowarski and M. P. Bloustein, *Nature (London)*, **300**, 650 (1982).

159. A. L. Boynton, W. L. McKeehan and J. F. Whitfield (eds), *Ions, Cell Proliferation and Cancer*, Academic Press, New York, 1982.
160. P. Laszlo, *Angew, Chem., Int. Ed. Engl.*, **17**, 254 (1978).
161. F. Inagaki and T. Miyazawa, *Prog. Nucl. Magn. Reson. Spectrosc.*, **14**, 67 (1981).
162. C. C. Bryden, C. N. Reilley and J. F. Desreux, *Anal. Chem.*, **53**, 1418 (1981).
163. M. M. Pike, S. R. Simon, J. R. Balschi and C. S. Springer, *Proc. Natl. Acad, Sci. USA*, **79**, 810 (1982).
164. M. M. Pike and C. S. Springer, *J. Magn. Reson.*, **46**, 348 (1982).
165. R. K. Gupta and P. Gupta, *J. Magn. Reson.*, **47**, 344 (1982).
166. J. W. Pettegrew, D. E. Woessner, N. J. Minshew and T. Glonek, *J. Magn. Reson.*, **57**, 185 (1984).
167. P. J. Brophy, M. K. Hayer and F. C. Riddel, *Biochem. J.*, **210**, 961 (1983).
168. Y. Boulanger, P. Vinay, M. T. Phan Viet, R. Guardo and M. Desroches, *Magn. Reson. Med.*, **2**, 495 (1985).
169. S. R. Gullans, M. J. Avison, T. Ogino, G. Giebisch and R. G. Shulman, *Am. J. Physiol.*, **249**, F160 (1985).
170. J-P Monti. P. Gallice, A. Crevat, M. El Mehdi, C. Durand and A. Murisasco, *Clin. Chem.*, **32**, 104 (1986).
171. A. M. Kumar, A. Spitzer and R. K. Gupta, *Kidney Int.*, **29**, 747 (1986).
172. B. M. Rayson and R. K. Gupta, *J. Biol. Chem.*, **260**, 7276 (1985).
173. P. S. Hubbard, *J. Chem. Phys.*, **53**, 985 (1970).
174. T. E. Bull, *J. Magn. Reson.*, **8**, 344 (1972).
175. T. E. Bull, J. Andrasko, E. Chioncone and S. Forsén, *J. Mol. Biol.*, **73**, 251 (1973).
176. A. Delville, C. Detellier and P. Laszlo, *J. Magn. Reson.*, **34**, 301 (1979)
177. T. Ogina, J. A. den Hollander and R. G. Shulman, *Proc. Natl. Acad. Sci. USA*, **80**, 1099 (1983).
178. A. M. Castle, R. M. Macnab and R. G. Shulman, *J. Biol. Chem.*, **261**, 3288 (1986).
179. M. M. Civan and M. Shporer. In L. J. Berliner and J. Reuben (eds), *Biological Magnetic Resonance*, Vol. 1, Plenum Press, New York, 1978, p. 1.
180. M. M. Pike, E. T. Fossel, T. W. Smith and C. S. Springer, *Am. J. Physiol.*, **246**, C528 (1984).
181. R. K. Gupta, A. B. Kostellow and G. A. Morrill, *J. Biol. Chem.*, **260**, 9203 (1985).
182. P. F. Renshaw, H. Blum and J. S. Leigh, *J. Magn. Reson.*, **69**, 523 (1986).
183. N. A. Matwiyoff, C. Gasparovic, R. Wenk, J. D. Wicks and A. Rath, *Magn. Reson. Med.*, **3**, 164 (1986).
184. J. Pekar and J. S. Leigh, *J. Magn. Reson.*, **69**, 582 (1986).
185. J. Pekar, P. F. Renshaw and J. S. Leigh, *J. Magn. Reson.*, **72**, 159 (1987).
186. A. Bax, *Two-Dimensional Nuclear Magnetic Resonance in Liquids*, Delft University Press, Delft, 1984.
187. G. A. Smith, P. G. Morris, T. R. Hesketh and J. C. Metcalfe, *Biochim. Biophys. Acta*, **889**, 72 (1986).
188. T. Ogino, G. I. Shulman, M. J. Avison, S. R. Gullans, J. A. den Hollander and R. G. Shulman, *Proc. Natl. Acad. Sci. USA*, **82**, 1099 (1985).
189. W. R. Adam, A. P. Koretsky and M. W. Weiner, *Biophys. J.*, **51**, 265 (1987).
190. K. R. Thulborn and J. J. H. Ackerman, *J. Magn. Reson.*, **55**, 357 (1983).
191. L. Garfinkel, R. A. Altschuld and D. Garfinkel, *J. Mol. Cell. Cardiol.*, **18**, 1003 (1986).
192. R. K. Gupta, J. L. Benovic and Z. B. Rose, *J. Biol. Chem.*, **253**, 6172 (1978).
193. K. V. Vabavada, B. D. Ray and B. D. Nageswara Rao, *J. Inorg. Biochem.*, **21**, 323 (1984).
194. S. T. Wu, W. M. Pieper, J. M. Salhory and R. S. Eliot, *Biochemistry*, **20**, 7399 (1981).

195. K. M. Brindle and I. D. Campbell. In T. L. James amd A. R. Margulis (eds) *Biomedical Magnetic Resonance*, San Francisco Radiology Research Foundation, San Francisco, 1984, p. 243.
196. R. J. Labotka, J. A. Warth, V. Winecki and A. Omachi, *Anal. Biochem.*, **147**, 75 (1985).
197. K. Kirk and P. W. Kuchel, *J. Magn. Reson.*, **68**, 311 (1986).
198. M. Cerdonio, S. Morante, D. Torresani, S. Vitale, A. DeYoung and R. W. Noble, *Proc. Natl. Acad. Sci. USA*, **82**, 102 (1985).
199. J. E. Peterson and R. D. Stewart, *Arch. Environ. Health*, **21**, 165 (1970).
200. A. Kalk and H. J. C. Berendsen, *J. Magn. Reson.*, **24**, 343 (1976).
201. D. L. Rabenstein, A. A. Isab and D. W. Brown, *J. Magn. Reson.*, **41**, 361 (1986).
202. K. Akasaka, M. Konrad and R. S. Goody, *FEBS Lett.*, **96**, 287 (1978).
203. B. D. Sykes, W. E. Hull and G. H. Snyder, *Biophys. J.*, **21**, 137 (1978).
204. F. F. Brown and I. D. Campbell, *Philos. Trans. R. Soc. London. Ser. B*, **289**, 395 (1980).
205. A. Daniels, R. J. P. Williams and P. E. Wright, *Nature (London)*, **261**, 321 (1976).
206. B. E. Chapman, D. I. Cook, J. Gerrard, A. P. Healey, P. W. Kuchel and J. A. Young, *J. Physiol.*, **330**, 36P (1982).
207. C. R. Middlehurst, G. R. Beilharz, G. E. Hunt, P. W. Kuchel and G. F. S. Johnson, *J. Neurochem.*, **42**, 878 (1984).
208. C. R. Middlehurst, G. F. King, G. R. Beilharz, G. E. Hunt, G. F. S. Johnson and P. W. Kuchel, *J. Neurochem.*, **43**, 1561 (1984).
209. K. Yoskizaki, Y. Seo and H. Nishikawa, *Biochim. Biophys. Acta*, **678**, 283 (1981).
210. C. Arús, M. Bárány, W. M. Westler and J. L. Markley, *FEBS Lett.*, **165**, 231 (1984).
211. C. Arús, M. Bárány, N. M. Westler and J. L. Markley, *Clin. Physiol. Biochem.*, **2**, 49 (1984).
212. K. Ugurbil, M. Petein, R. Maidan, S. Michurski, J. N. Cohn and A. H. From, *FEBS Lett.*, **167**, 73 (1984).
213. P. F. Agris and I. D. Campbell, *Science*, **216**, 1325 (1982).
214. Z. H. Endre and P. W. Kuchel, *Kidney Int.*, **28**, 6 (1985).
215. D. L. Rabenstein, *Anal. Chem.*, **50**, 1265A (1978).
216. D. L. Rabenstein and T. T. Nakashima, *Anal. Chem.*, **51**, 1465A (1979).
217. K. M. Brindle, F. F. Brown, I. D. Campbell, D. L. Foxall and R. J. Simpson, *Biochem. J.*, **202**, 589 (1982).
218. G. H. Weiss and J. A. Feretti, *J. Magn. Reson.*, **55**, 397 (1983).
219. P. W. Kuchel, B. T. Bulliman, B. E. Chapman and G. L. Mendz, *J. Magn. Reson.*, **76**, 136 (1987).
220. P. W. Kuchel, B. S. Singh, G. E. Hunt, G. F. S. Johnson, W. B. Begg and A. J. Jones, *N. Engl. J. Med.*, **303**, 705 (1980).
221. A. J. Jones and P. W. Kuchel, *Clin. Chim. Acta*, **104**, 77 (1980).
222. P. W. Kuchel, G. E. Hunt, G. F. S. Johnson, G. R. Beilharz, B. E. Chapman, A. J. Jones and B. S. Singh, *J. Affective Disorders*, **6**, 83 (1984).
223. G. R. Beilharz, C. R. Middlehurst, P. W. Kuchel, G. E. Hunt and G. F. S. Johnson, *Anal. Biochem.*, **137**, 324 (1984).
224. B. Thiault and M. Mersseman, *Org. Magn. Reson.*, **7**, 575 (1975).
225. B. E. Chapman, G. R. Beilharz, M. J. York and P. W. Kuchel, *Biochem. Biophys. Res. Commun.*, **105**, 1280 (1982).
226. E. F. Domino, R. R. Sharp, S. Lipper, C. L. Ballast, B. Delidow and M. R. Bronzo, *Biol. Psychiatry*, **20**, 1277 (1985).
227. E. F. Domino, *Biol. Psychiatry*, **22**, 396 (1987).
228. P. W. Kuchel and B. E. Chapman, *Biomed. Biochim. Acta*, **9**, 1137 (1983).

229. J. I. Vandenberg, G. F. King and P. W. Kuchel, *Arch. Biochem. Biophys*, **242**, 515 (1985).

230. J. I. Vandenberg, P. W. Kuchel and G. F. King, *Anal. Biochem.*, **155**, 38 (1986).

231. G. F. King, C. R. Middlehurst and P. W. Kuchel, *Biochemistry*, **25**, 1054 (1986).

232. J. R. Bales, J. K. Nicholson and P. J. Sadler, *Clin. Chem.*, **31**, 757 (1985).

233. J. K. Nicholson, M. P. O'Flynn, P. J. Sadler, A. F. Macleod, S. M. Juul and P. Sönksen, *Biochem. J.*, **217** 365 (1984).

234. J. R. Bales, J. D. Bell, J. K. Nicholson and P. J. Sadler, *Magn. Reson. Med.*, **3**, 849 (1986).

235. R. A. Iles, A. J. Hind and R. A. Chalmers, *Clin. Chem.*, **31**, 1795 (1985).

236. S. Yamaguchi, N. Koda, Y. Eto and K. Aoki, *J. Pediatr.* **106**, 620 (1985).

237. W. Lehnert, D. Hunkler, *J. Pediatr.*, **108**, 166 (1986).

238. J. K. Nicholson, P. J. Sadler, J. R. Bales, S. M. Juul, A. F. Macleod and P. H. Sönksen, *Lancet*, **29**, 751 (1984).

239. R. A. Iles, M. J. Buckingham and G. E. Hawkes, *Biochem. Soc. Trans.*, **11**, 374 (1983).

240. M. D. Coleman and R. S. Norton, *Xenobiotica*, **16**, 69 (1986).

241. J. K. Nicholson, J. A. Timbrell, D. P. Higham and P. J. Sadler, *Hum. Toxicol.*, **3**, 334 (1984).

242. J. K. Nicholson, J. A. Timbrell and P. J. Sadler, *Mol. Pharmacol.*, **27**, 644 (1985).

243. J. L. Bock, *Clin. Chem.*, **28**, 1873 (1982).

244. J. K. Nicholson, M. J. Buckingham and P. J. Sadler, *Biochem. J.*, **211**, 605 (1983).

245. J. D. Bell, J. C. C. Brown and P. J. Sadler, *Chemistry in Britain*, **24**, 1021 (1988).

246. J. D. Bell, J. C. C. Brown, J. K. Nicholson and P. J. Sadler, *FEBS Lett.*, **215**, 311 (1987).

247. M. Traube, J. L. Bock and J. L. Boyer, *Ann. Intern. Med.*, **98**, 171 (1983).

248. O. A. C. Petroff, R. K. Yu and T. Ogino, *J. Neurochem.*, **47**, 1270 (1986).

249. J. D. Bell, J. C. C. Brown, P. J. Sadler, A. F. MacLeod, P. H. Sönksen, R. D. Hughes and R. Williams, *Clin. Sci.*, **72**, 563 (1987).

250. E. T. Fossel, J. M. Carr and J. McDonagh, *N. Engl. J. Med.*, **315**, 1369 (1986).

251. P. G. Williams, M. A. Helger, L. C. Wright, M. Dyne, R. M. Fox, K. T. Holmes, G. L. May and C. E. Mountford, *FEBS Lett.*, **192**, 159 (1985).

252. K. T. Holmes, P. G. Williams, G. L. May, P. Gregory, L. C. Wright, M. Dyne and C. E. Mountford, *FEBS Lett.*, **202**, 122 (1986).

253. M. Bloom, K. T. Holmes, C. E. Mountford and P. G. Williams, *J. Magn. Reson.*, **69**, 73 (1986).

254. J. A. den Hollander, K. Ugurbil, T. R. Brown, M. Bednar, C. Redfield and R. G. Shulman, *Biochemistry*, **25**, 203 (1986).

255. J. A. den Hollander, K. Ugurbil and R. G. Shulman, *Biochemistry*, **25**, 212 (1986).

256. M. E. Stromski, F. Arias-Mendoza, J. R. Alger and R. G. Shulman, *Magn. Reson. Med.*, **3**, 24 (1986).

257. J. R. Alger, K. L. Behar, D. L. Rothman and R. G. Shulman, *J. Magn. Reson.*, **56**, 334 (1984).

258. P. Canioni, J. R. Alger and R. G. Shulman, *Biochemistry*, **22**, 4974 (1983).

Chapter 7

Automatic NMR Analysis

M. Spraul

Bruker Analytische Messtechnik GmBH, Karlsruhe, Federal Republic of Germany

R.-D. Reinhardt

BASF AG, Central Research, Ludwigshafen, Federal Republic of Germany

1   DEVELOPMENT OF AUTOMATION IN NMR

During the past three decades NMR has become an invaluable analytical method. Very different systems ranging from proteins, organometallics and polymers to natural and synthetic organic compounds are routinely analysed by NMR today. Analytical NMR has also gained in importance in the investigation of chemical reactions and also has developed important environmental applications.

With the increasing capability of the technique, the number of samples which had to be examined by NMR began to rise dramatically in about the mid-1960s. At that time efforts to automate both NMR spectrometers and spectral interpretation were initiated. In this chapter we give a summary of how automation in NMR was developed and show how routine automatic analysis can be performed on modern spectrometers. The chapter closes with a preview of what can be expected in the future.

The analysis of a sample by NMR can be divided into 6 six major steps:

(i) the sample must be brought to the laboratory:
(ii) the sample has to be prepared;
(iii) the appropriate spectrum must be recorded;
(iv) the spectrum must be interpreted.

During the interpretation it may become necessary to repeat one or more of steps (i) to (iv) again.

(v) If interpretation comes to a conclusive result this has to be documented;
(vi) the costs of the analysis must be charged to some account. This step is very important in industry.

The handling of steps (i), (v) and (vi) is a problem of laboratory organization and can be done by laboratory management systems. If steps (iii)–(v) are fully automated this will increase the sample throughput of a spectrometer dramatically because the system can then be run unattended overnight and at weekends. Consequently, the total cost per sample will decrease and the protection of costs due to staffing can be reduced.

Efforts to automate steps (iv) and (v) began in 1970. In the central research laboratory of BASF, a Varian HR220 superconducting NMR CW spectrometer (Figure 1) was the first to be equipped with a laboratory-built sample changer with the capacity for 24 samples. A computer with 8 kbyte of memory was used to control the spectrometer which had no lock device.

In this spectrometer, after a starting signal the probe head was removed from the magnet and positioned directly under a sample. The sample tubes were stored in a sample changer which was designed like a flat disk. A pneumatic device lowered the sample into the probe head and rotation was checked by means of a reflected light beam. If rotation was correct the probe head was repositioned in the magnet. Two gradients (z and z^2) were shimmed automatically using the lineshape of the TMS signal as a shim criterion. After the adjustment of homogeneity was complete, the spectrum was recorded and the TMS signal referenced to 0 ppm. The largest signal in the spectrum was scaled such that it filled the available paper during the plot. An integral was computed and plotted. After the plot was finished on a normal x,y-plotter, the spectrum was stamped with the sample number and transferred to a 'paper basket'. The probe head was removed from the magnet, the sample returned to its position in the sample changer and the changer moved to the next position ready for the next experiment. It was possible to measure 24 samples overnight and the same number during the day, thus doubling the spectrometer's sample throughput.

Fig. 1. The old Varian HR220 superconducting CW spectrometer. It was the first spectrometer to be automated in the central laboratory at BASF in 1970

The HR220 was used until 1977, when it was taken out of service because of high helium consumption.

In 1977 a 360 MHz FT spectrometer was purchased and between 1980 and 1984 a new sample changer and the necessary software and hardware were developed in close collaboration with Bruker. The project was sponsored by the German Ministry of Research and Development. Bruker constructed all the software components and hardware devices, including the autoshim, autolock and automatic receiver gain adjustment. The sample changer itself was constructed at BASF. The concept of the sample changer was totally different from that of the earlier device and it allowed storage of up to 60 samples. The samples were lifted from the magnet by air pressure and placed in the sample changer with pneumatic device. Five prototypes of the sample changer were produced at BASF and three of them are still in use. One was given to Bruker for further development and the construction of the modern commercial version of automatic sample changer. Today every manufacturer of NMR spectrometers offers sample changers which are based on a variety of concepts. Currently, most industrial NMR laboratories employ sample changers and an increasing number of universities and research laboratories are performing their routine NMR analyses with the aid of sample changers[2]. Today's spectrometers are fully automated and allow measurement of more than 100 samples in one run without operator intervention. The extent to which NMR automation has been developed will be demonstrated in Section 3.

2 AUTOMATIC SAMPLE PREPARATION

The last step to be automated was that involving preparation of the samples. In order to have standardized conditions and obtain comparable routine spectra, the samples should ideally be prepared in the NMR laboratory itself. In our laboratory (BASF)[3] we started to develop a sample preparation system for this purpose in 1985, based on a Zymark robot. The system has the capacity to prepare 105 NMR samples and a picture of the entire system (in use since April 1986) is presented in Figure 2. The system can handle both liquids and solids and prepare samples in three different solvents.

The samples arrive in the laboratory in standard vials with screw-caps. They are placed in three racks, each of which can hold a maximum of 35 vials. Normally liquids and solids are placed in different racks (a in Figure 2) and all

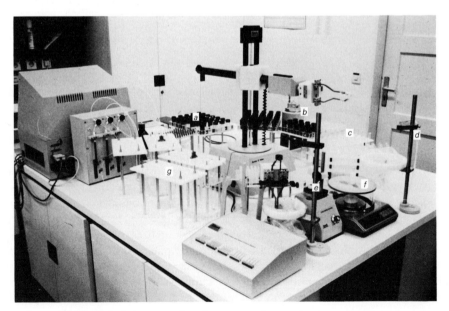

Fig. 2. Laboratory robot for automatic sample preparation

compounds which require the same solvent are grouped together. For each samples there must be a minimum of 500 mg for solids and 0.5 ml for liquids. The robot is started by placing its hand over three edges of each rack. The coordinates of a sample in a rack are then computed by the data system connected to the robot. All other actions which the robot has to perform are programmed before the experiment. In a normal run, the operator only has to enter those places in the rack occupied by samples, and indicate which of the samples are solid, which are liquids and which solvent is required for each sample.

After the system has been started the robot takes the first vial and places it in a 'capper-station' (b in Figure 2). The hand of the robot holds the cap while the vial is turned to uncap it. After the cap has been removed it is replaced on a small board in a defined position. Now the procedure diverges, depending on whether the sample is a liquid or a solid.

If the sample is a liquid it is transferred by a syringe to a test-tube (c in Figure 2). The robot takes the test-tube and moves it to the solvent station (d in Figure 2) to dilute the sample with the predefined solvent. In the next step the test tube is rotated (e in Figure 2) to mix the sample and solvent.

If the sample is a solid, the robot places a test-tube in a small plastic block on a balance (f in Figure 2). The balance is coupled to the data station by a normal RS232 interface. The robot changes its hand to a vibrotor hand, takes the sample vial, tilts it by a defined angle and moves it over the test-tube (Figure 3). By slowly increasing the angle of the sample vial and the vibration rate, the solid will fall into the test-tube. If the balance detects that 30–50 mg of the sample is in the test-tube, the sample vial is placed back into the capper station.

For solid samples, a problem arises with the particle size. If there are large particles, the required amount of sample in the test-tube might suddenly become much to high. On the other hand, if the solid is a very fine powder, it might stick to the walls of the glass vial because of electrostatic effects and nothing will fall into the test-tube. In either case the preparation of the sample is aborted, the sample is returned to the rack, an error message is displayed on a printer and the preparation of the next sample is started. If no problems are detected, the robot changes its hand again and the test-tube is moved to the solvent station, diluted and rotated to dissolve the solid.

While the test-tube is rotating, the robot caps the sample vial and returns it to the rack. The solution is then transferred to the NMR tube with a syringe. The empty NMR tubes wait in a rack (g in Figure 2) and above the NMR tubes a plate with holes helps to guide the syringe into the NMR tube correctly. After all samples have been prepared the plate is removed and the NMR tubes are capped. They can be loaded into the spectrometer's automatic sample changer, ready to be examined.

In the future, one can imagine that the sample tubes brought to the laboratory may be labelled with a bar code which can be read by the robot. The bar code

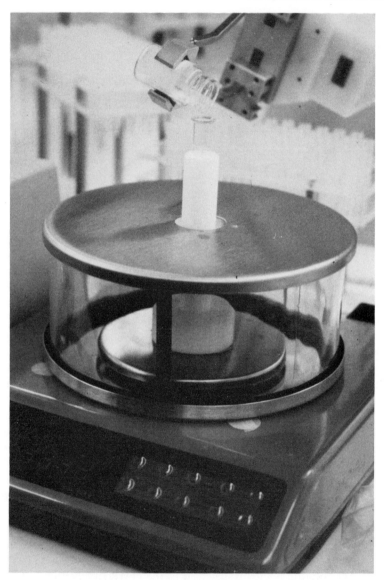

Fig. 3. The laboratory robot weighting a solid sample

will contain information about the sample and the experiment to be executed. The robot will decide whether the sample is solid or liquid and choose the correct solvent from the information contained in the bar code. The robot will create a new bar code which can be read by the spectrometer's bar code reader and fix it to the NMR tube. When all this is possible, the only manual steps which remain are to place the samples in the racks and to transfer the NMR tubes to the sample changer. In such a totally automated system, the responsibility falls heavily on the chemist to give correct information about the sample and to select correctly the experiments to be performed. Otherwise, a sample with a very time-consuming experiment might 'automatically' prevent or delay execution of the measurements on all following samples.

3 AUTOMATIC ACQUISITION AND PROCESSING OF NMR SPECTRA

3.1 Defining sample conditions

A modern NMR sample changer has up to 120 holders. Therefore, the efficiency with which sample conditions can be defined is one criterion for an effective automation software package. The more questions to be answered or the more parameters to be set, the more time consuming the set-up will be. Two different groups of users exist today: (a) those running only routine one-dimensional experiments and (b) those who wish to incorporate the most advanced two-dimensional sequences. Both groups of users have different expectations of the spectrometer set-up. The first group needs as little interaction as possible, whereas the second demands access to most of the spectrometer parameters. A dialogue type of software is acknowledged by most users of automated NMR systems to be the optimum approach for the set-up.

The group (a) user defined above needs the input to be as short and simple as possible in order to allow a large number of samples to be processed in a short time. For the acquisition of simple one-dimensional proton spectra several laboratories in various parts of the world have a throughput of more than 200 samples per day.

The minimum information input for a sample must include (i) the experiment, (ii) the solvent and (iii) a sample code or title, where standard parameters for the experiment–solvent combinations are used.

When using standard parameters, a group (b) user requires the option to change parameters selectively, and this should be possible via an interactive dialogue. One possible way to fulfil the above needs is to define standard experiments for all types of measurements which are commonly executed in a laboratory. The parameters can then be used without change or in a modified form for each specific sample.

If each sample needs its own specific measurements, a dialogue asking for more details is more useful. For example, the dialogue could ask for acquisition type, processing type and plot options. A typical experiment set-up dialogue is shown in Figure 4.

Enter four character code:

Sample number:

Experiment:

Solvent:

Enter plot title:

Change any more parameters?

Perform another experiment with the same sample?

Set up another sample?

Fig. 4. Typical set-up for sample-changer measurements

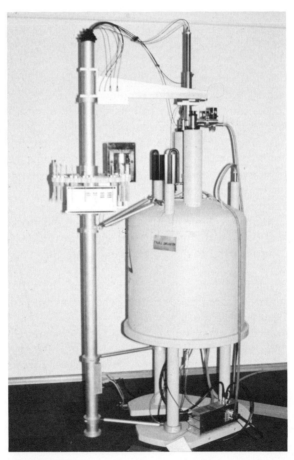

Fig. 5. Automatic sample changer equipped with bar code reader mounted on a 250 MHz magnet

As already mentioned (Section 2), another approach to setting up the parameters is by use of a bar code. Figure 5 shows a sample changer equipped with bar code reader (Figure 6). In this system, the bar code information is mounted on a collar on top of the spinner. The bar code reader grasps the sample in the holder and spins it to read the code.

Bar codes can be used in two ways: (i) bar code number only for sample identification or (ii) bar code number + measurement conditions coded. In a sample changer without a bar code reader, the holder position identifies the sample and therefore the programmed measurement conditions. If by mistake tubes are exchanged, identification of the correct spectra is difficult. The bar

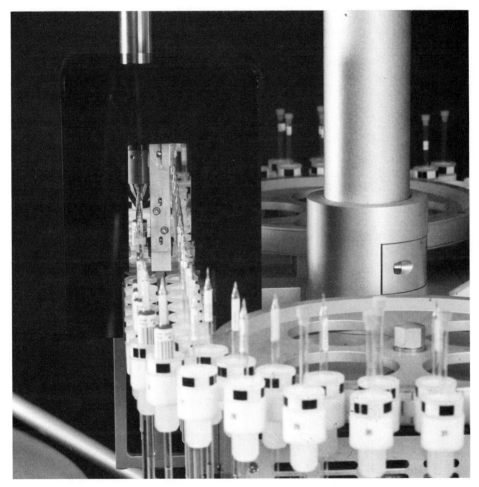

Fig. 6. Bar code reader attached to a sample changer with 120 holders

code approach is superior because the sample identification is independent of the position sample in the holder.

If the bar code label contains only the bar code number, a manual parameter set-up is still required. For each bar code number, the experiments must be specified in a manner similar to the common dialogue set-up. A bar code label can also contain the information to describe completely the sample and experiments to be performed. No manual interaction is then necessary. The label must contain the following information: (i) a label number; (ii) an experiment descriptor (standard experiments); (iii) the solvent to be used and (iv) a user identification code.

A commercially available system has 12 digits, which are divided as follows: (i) label number, 5 digits; (ii) experiment, 2 digits; (iii) solvent code, 2 digits; and (iv) user identification code, 3 digits.

A label printer driven by the NMR system's host computer or a personal computer is offered for that bar code system. The use of a bar code also avoids the possibility of double measurement on samples which may be forgotten in the sample changer when starting a new series. Double measurement is precluded by a protocol file on the system disk, which contains an entry for each label number measured.

Figure 7 shows a bar code label, also containing the coded information in user-readable form. As can be seen from the full bar code system, no manual

Fig. 7. Bar code for automatic NMR spectroscopy. Apart from the bar code itself, the experiment, solvent, user and label number are printed in user-readable form

interaction is necessary, so when using such a system the software must be able to optimize automatically the measurement conditions for an unknown sample. Options available in this respect are explained in Section 3.3.

3.2 Important sample-changer features

A simple one-dimensional proton measurement in an automated spectrometer including sample changing and shimming takes about 4–5 min. This means that, in principle, 120 samples can be measured in 10 h (which can readily be performed overnight). Clearly, the number of sample holders should be > 100.

Temperature-controlled holders are needed, for example, for polymer measurement. To avoid long waiting periods where the sample is molten in the magnet and has to be homogenized, it is preferable to use holders which can be preheated. Sometimes the opposite is needed, i.e. to have sample holders which are cooled. This can be achieved, for example, by using Peltier elements. Figure 8 shows a heated holder system (for the Bruker sample changer) where each holder can be maintained at a different temperature. For light-sensitive samples, closed sample holders must be used.

The interface between sample changer microprocessor and NMR system computer should allow information transfer in both directions, so that the sample position or any error messages can be reported back to the system computer.

3.3 Steps needed in an automatic acquisition

(a) Sample change

When starting a new sample measurement, the fact that no sample is already in the magnet must be verified before the next sample is loaded. To check this, the following tests can be performed: (i) try to autolock the spectrometer; (ii) try to spin the sample; (iii) try to blow out the sample and detect it on top of the magnet. If all three tests fail, then the probe is vacant and the sample can be inserted into the magnet.

(b) Start sample rotation

(c) Autolock

If the lock parameters for the solvent are known to the system, then lock field and lock power can be adjusted before starting the autolock. The system should be able to increase the lock sweep width, power and gain if the first trail fails.

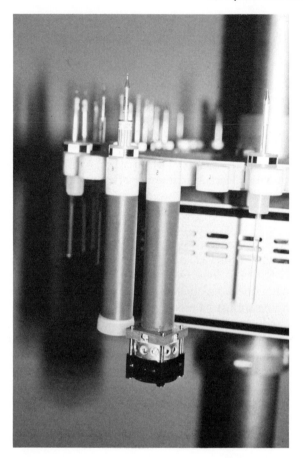

Fig. 8. Heatable holder for an automatic sample changer. The temperature is adjusted at the bottom of the holder. Electric power is supplied by a collecting ring

(d) Computer shim

All shims should be accessible by the computer shim procedure, although for most applications optimizing z^1 and z^2 is sufficient to establish good homogeneity. Complete shim value sets corresponding, for example, to different filling heights of the NMR tubes must be available. Computer shimming (using either the lock level or the FID area as criteria) is necessary in automatic spectral acquisition. Sometimes, shimming on the FID in an unlocked mode is necessary; for example, when deuterium spectra are acquired, then the computer shim can optimize on the proton FID area.

(e)   Autoshim

After finishing the computer shim the system should switch to the autoshim procedure to maintain the optimum homogeneity throughout the spectral acquisition.

(f)   Tuning and matching adjustment of the probe for different nuclei

For Bruker NMR systems, a probe head which can be automatically optimized for the measurement of four nuclei is available. Tuning and matching of the probe are adjusted pneumatically for the different frequencies using positions coded into the parameter sets for certain X-nuclei. ^{1}H is measured on the decoupler coil. Such an arrangement allows the automatic measurement of ^{1}H, ^{19}F, ^{31}P and ^{13}C within a sample changer run and run and needs no manual intervention.

(g)   Adjustment of receiver gain

To fill the digitizer completely, the receiver gain must be optimized. This procedure is also needed for all one- and two-dimensional pulse sequences used in automation and should be a basic part of the acquisition program.

For example, when executing a DEPT sequence[4] (DEPT = distortionless enhancement by polarization transfer) on a sample of low concentration, the solvent line (dominating the FID of a normal broad-band decoupled acquisition) will disappear and a higher receiver gain can be used. It is obvious that the receiver gain adjustment has to be executed on the pulse sequence actually defined for acquisition.

(h)   Adjustment of signal-to-noise ratio

In a standard acquisition, the number of transients is predefined and might not fulfil the requirement for a good (S/N) signal-to-noise ratio for the peaks of interest.

Criteria to allow automatic setting of the number of transients need to be defined. The first solution offered was simply to specify an amplitude ratio between the first and last data points of the FID. This, of course, does not solve the problem when the solvent or a reference signal dominates the FID.

An alternative solution is to define specific regions of the spectrum where the signal and the noise are to be measured. Then, despite the presence of any large peaks in the spectrum the S/N is optimized on the largest peak in the area specified. An example demonstrates the necessity of this feature: when ^{13}C NMR is used to analyse branching in polyethylene, the interesting peaks are very small, while the spectrum is dominated by a large peak due to the main-chain carbons. The way to proceed is to collect a certain number of scans x, then transform and calculate the S/N in the region of interest. If the necessary S/N is achieved, then stop; otherwise, add another x scans and check again, repeat the procedure until a satisfactory S/N is reached. A safety check is necessary in this scheme to avoid an endless run if there is no peak in the specified region. The check can be done by simply defining a maximum number of loops through x scans.

An improvement in this feature is to transfer scan number determinations to subsequent experiments with the same sample. An example will illustrate this: consider an automatic run of DEPT and a 2D C/H correlation[5] on an unknown sample. DEPT is acquired first using the S/N criteria described above and the scan number is determined and stored. An optimum scan number for the 2D experiment can be calculated by using a scan number conversion table containing (i) the experiment where the number of scans necessary to reach a specified S/N is determined; (ii) the nature of the experiment which follows; (iii) a multiplication factor for the scan number; and (iv) the minimum phase cycle for the experiment which follows.

The conversion factor from DEPT to 2D CH correlation could be set to 0.1, the minimum phase cycling for the 2D experiment is 4. Therefore, if 64 scans were needed for the DEPT to reach the defined S/N, 6.4 would be the scan number for the 2D; as the minimum phase cycle is 4 the scan number would be adjusted to 8.

(i)   Optimizing the sweep width

In 2D experiments, where digital resolution is limited, it is desirable to optimize the spectral window automatically. The same is true for measuring the spectra of X-nuclei where a wide spectral range is possible, e.g. ^{31}P or ^{19}F.

A possible approach is as follows:

(i)   first take an overview spectrum with a large spectral width;
(ii) transform, correct the phase and baseline;
(iii) automatically calculate integral reset points;
(iv) take the highest and lowest field regions found and calculate the new sweep width accordingly;
(v)   recalculate 2D parameters depending on the experiment;
(vi) acquire spectrum with optimized limits.

Some precautions in automatically adjusting sweep width should be taken into account by 'intelligent' software. These include the following:

(i)   Maintaining the recycle time at a constant value compared with standard parameters (recycle time = acquisition time + relaxation delay). When the sweep width is reduced, the acquisition time increases: halving the sweep width doubles the acquisition time. For example, a 2D COSY[6] experiment set for 2 h duration could easily take 8 h if the sweep width was reduced without compensation. This could destroy an entire overnight measuring schedule. The relaxation delay should be decreased by the same amount as the acquisition time is increased. If the relaxation delay cannot be reduced further, the time domain of the FID must be reduced.

(ii) Allow the exclusion of certain spectral regions. If the solvent or the reference is the lowest or highest field signal, it might be necessary to exclude these peaks from the optimized sweep width. This problem can be solved by maintaining a file on the system disk which contains the specific regions of no interest for each solvent. It is also desirable to ensure that folded peaks do not fall into regions where peaks are found.

(iii) Increase the sweep width found. To avoid baseline distortions caused by peaks lying close to the edges of the spectral window, the optimum sweep width found should be increased. The following values give acceptable results: ^{1}H spectra increase by 10% of the sweep width on both sides; and X-nuclei increase by 5% on both sides.

(iv) Determination of the sensitivity of integral region search. If only the main compound in a mixture is needed to optimize an experiment, the system need only determine regions showing large integrals. However, if minor compounds in a mixture need to be analysed, small integrals must also be recognized. This can be achieved by defining a certain fraction of the largest integral as the threshold level below which integrals are not recognized.

(v) For heteronuclear 2D correlations, the system must be able to optimise both proton and X-nucleus chemical shift ranges.

A standard 2D CH correlation requires that the system first performs a proton overview to obtain integral regions for optimizing the proton domain. Then the system performs a DEPT experiment with S/N determination and obtains integral regions to optimize the carbon domain. Thirdly, the 2D correlation experiment (with optimized sweep ranges in both dimensions as well as optimized scan numbers) is executed. Finally, the data must be processed and plotted. Such an experiment can be fully specified by the bar code on a submitted sample and executed without operator intervention.

3.4 Automatic data processing

Automatic processing includes all software manipulations from transformation of the FID (free induction decay) to the final plot or peak print-out.

After window multiplication, Fourier transformation, phase correction and baseline correction, the following steps are useful.

(a) Create integral resets

Integral resets are allowed whenever the baseline is reached after a peak has been identified, usually using the second derivative of the spectrum. An integral threshold [as described in Section 3.3 i)] is needed to suppress small integrals. Other parameters which influence automatic integration are (i) the minimum distance between two lines to allow an integral reset, which is necessary to avoid resets in well resolved multiplets, and (ii) prolongation of the machine-designated integral limits; automatic resets may be too close to the signals and for visual inspection afterwards it may be desirable to prolong or extend the integral curves.

(b) Rescaling of chemical shift values

Depending on concentration, pH, etc., chemical shifts may vary within different samples. If the absolute frequency of the solvent and the shift difference to a reference standard is known, the system can automatically recalibrate the chemical shift axis. This procedure can also be used to identify the solvent and reference resonances (TMS), which are needed for optimum plotting [see Section 3.4 (c)].

(c) Plotting

Plotting with or without integrals is available. Y-scaling is performed by disregarding the solvent and reference peaks if they are the largest signals in the spectrum. Y-scaling should always yield the largest solute peak at full-scale. When fixed plots limits are used, (i.e. not covering the full range acquired), the system should be able to perform additional plots, if peaks are found in the regions which are not plotted. Peak labeling on the plot should be in either Hz or ppm.

(d) Automatically expanded plots

A reasonable approach is to expand the regions between the integral resets with a user definable scale ($Hz\,cm^{-1}$ or $ppm\,cm^{-1}$). By this method, peaks and integrals can be expanded to full-scale with the appropriate scaling factors listed on the plot. Alternatively, to avoid a waste of paper, regions of interest can be pre-defined and all peaks falling outside these regions are not expanded.

(e) Peak picking

Peak picking is needed in order to generate peak lists or to label peaks on the plot. Defining a minimum intensity is not sufficient to avoid noise being labelled in addition to 'real' signals. The best approach is to define a peak-picking constant, which is a multiple of the noise level; if that constant is set larger than 1, only peaks bigger than the noise are labelled. By further increasing this value only major peaks can be selected and listed.

(f) Integral-based calculations

For standard situations, calculations based on the integrals in certain regions (e.g. determining relative concentrations of components in a mixture) can be performed. This can be used in quality control.

(g) 2D Processing

Processing two-dimensional data includes the following:

- (i) window multiplication;
- (ii) 2D transformation;
- (iii) symmetry operations (if required), depending on the type of matrix;
- (iv) rescaling in both dimensions;
- (v) 2D noise calculation;
- (vi) creating a height level list for contour plotting, where the basic contour levels can be preset by the operator if wanted;
- (vii) 2D plot with the original (1D) spectra on both axes;
- (viii) 2D peak picking to identify scalar or dipolar interactions. One approach is to perform peak picking first in the F_2 dimension and then in the corresponding columns.

Figure 9 shows the result of a 2D peak picking and three automatically extracted colums of a CH correlation on a cyclic tripeptide.

3.5 A useful example of automatic spectrum interpretation

An automatic carbon multiplicity analysis is available, based on the DEPT experiment. Three DEPT spectra (DEPT 45, 90 and 135) and a broad-band decoupled ^{13}C spectrum must be acquired and processed. A comparison of the peak picking on these four files with regard to peak intensity and sign yields an output which contains the multiplicity of the carbon signals. The output can be presented in tabular form or as the labelling of the peaks on the plot. To ensure safe assignment, it is not possible to rely solely on complete elimination of peaks in subspectra (e.g. CH_2 and CH_3 in DEPT 90), since perfect nulling of signals is not always achieved in automatic spectroscopy. The reduction of peak intensities to a specified fraction is a better criterion.

If a multiplicity analysis is used as input to an automatic carbon library search, solvent signals must be excluded when transferring the peak list to the library. This can be achieved by pattern comparison of the pure solvent with the acquired spectrum (performed via the integral peak pickings). The result of the library search can be plotted on the spectrum or on a separate page following the plot.

SPLD3102.SMX   1
Min. Intensity = 0.0   Maxy = 18.00000   PP Constant = 3.00000
Intens. Level = 0.070   Noise = 0.10544   Sens. Level = 1.26524

#	Col.	Frequency F2 [HZ]	[PPM]	#	Row	Frequency F1 [HZ]	[PPM]	Intensity
1	46	9774.654	129.5195	1	13	2248.555	7.4919	1.467
				2	54	2163.689	7.2091	8.437
				3	59	2153.012	7.1735	8.320
				4	72	2126.153	7.0840	9.286
				5	77	2116.755	7.0527	18.342
				6	81	2106.833	7.0197	9.690
				7	89	2091.547	6.9687	1.853
				8	144	1977.152	6.5876	1.457
2	54	9704.514	128.5901	1	58	2155.221	7.1809	1.855
				2	77	2115.136	7.0473	6.614
				3	82	2106.104	7.0172	12.805
				4	147	1971.006	6.5671	1.364
3	655	4435.239	58.7693	1	442	1359.462	4.5295	12.967
				2	447	1349.872	4.4976	12.964
4	684	4180.981	55.4003	1	460	1322.206	4.4054	7.377
				2	464	1313.554	4.3766	13.760
				3	469	1304.587	4.3467	5.896
5	702	4023.166	53.3091	1	404	1439.598	4.7965	1.451
				2	465	1313.158	4.3752	8.950
				3	474	1294.577	4.3133	12.802
				4	522	1193.783	3.9775	12.599
				5	531	1175.311	3.9160	9.210
6	712	3935.491	52.1474	1	415	1415.827	4.7173	9.808
				2	422	1401.980	4.6712	11.782
				3	505	1229.138	4.0953	11.220
				4	512	1215.329	4.0493	10.022
7	742	3672.465	48.6622	1	575	1085.227	3.6158	7.541
				2	579	1075.991	3.5850	10.743
				3	583	1067.057	3.5553	8.706
8	751	3593.558	47.6166	1	558	1120.201	3.7323	4.096
				2	565	1105.580	3.6836	5.091
				3	632	965.419	3.2166	4.126
9	891	2366.106	31.3522	1	764	693.828	2.3117	2.606
				2	869	475.162	1.5832	1.913
10	905	2243.361	29.7258	1	741	740.743	2.4680	2.911
				2	953	300.842	1.0024	2.137
11	968	1691.007	22.4068	1	854	506.780	1.6885	1.630
				2	864	485.701	1.6183	1.764
				3	924	361.573	1.2047	2.292
				4	931	347.123	1.1566	2.032
12	981	1577.030	20.8965	1	837	541.187	1.8032	2.143
				2	914	382.174	1.2733	1.971

(a)

Fig. 9. (a) 2D peak-picking of a CH correlation experiment on a cyclic tripeptide. For each signal found in the carbon dimension the corresponding proton shifts are listed. (b) Plot of relevant columns automatically extracted from the 2D CH correlation shown in (a)

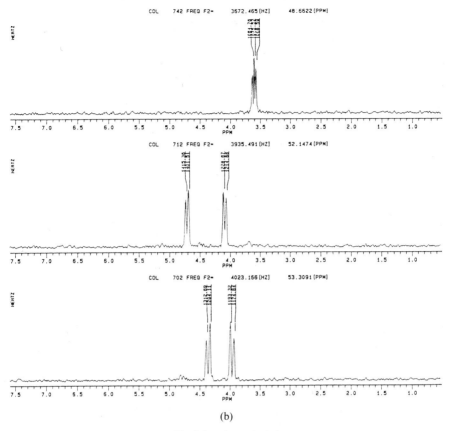

(b)

Fig. 9 (*caption opposite*)

Figure 10 shows the result of a multiplicity analysis on a cyclic tripeptide spectrum recorded on a 300 MHz spectrometer.

4 AUTOMATIC INTERPRETATION OF 1D SPECTRA

Automatic acquisition and processing of NMR spectra allow the measurement of significantly more spectra per day than was previously possible. In the BASF main laboratory, up to 300 proton NMR spectra and 100 carbon spectra are measured routinely every day and over the weekend on five superconducting, fully automated spectrometers. All carbon spectra and about 100 proton spectra are interpreted and documented daily. The remainder of the proton spectra are despatched to the chemists immediately after they have been recorded. The

SP310102.SER   4 SP310102.SER   4

Min. Intensity = 0.0	Maxy = 18.00000	PP Constant = 1.60000
Intens. Level = 0.070	Noise = 0.10796	Sens. Level = 0.69096
F1 = 15848.54	HZ = 210.0018 PPM	F2 = −377.20
	HZ = −4.9982 PPM	

#	Cursor	Frequency	PPM	Intensity	MUL
1	19743	5852.342	77.5467	2.279	
2	19800	5820.277	77.1218	2.399	
3	19857	5788.224	76.6971	2.365	
4	20568	5386.424	71.3731	15.493	D
5	23411	3779.716	50.0833	18.064	D
6	24080	3401.423	45.0707	15.032	T
7	25479	2610.998	34.5972	15.931	T
8	25869	2390.709	31.6782	17.327	D
9	26664	1941.340	25.7238	16.559	D
10	27007	1747.317	23.1529	14.713	T
11	27133	1676.166	22.2101	14.867	Q
12	27291	1586.828	21.0263	14.383	Q
13	27955	1211.517	16.0533	15.357	Q

Fig. 10. Automatic multiplicity analysis on a cyclic tripeptide (mul = multiplicity)

number of samples is increasing. One can obviously measure more spectra by using more spectrometers, but this creates the problem of interpreting an ever increasing amount of data. There are only two ways of solving this problem. The first choice might be to have more people to do the work, but this is a very expensive solution. The cheaper and perhaps more effective way to cope with the problem is to use modern computers for intelligent automatic spectrum interpretation.

Automatic spectrum interpretation starts on the spectrometer itself. Until such time as direct network links between the NMR spectrometer and a mainframe computer exist, it is very important to produce the spectral information in a user-friendly way. Lists of the chemical shifts of the signals observed together with their multiplicity information assist in simplifying the input of spectral data to a computer. If there is a local area network (LAN) between the spectrometer and the mainframe computer, data lists can be entered directly to an automatic interpretation program.

As an example of how the BASF SPECINFO program is used as a tool for routine spectrum interpretation, Figure 11 shows a typical carbon spectrum obtained in an automatic run. On the spectrum all lines are marked and their chemical shift and multiplicity are indicated. The suggested structure is that of menthol(**1**).

There are two possibilities for obtaining help from SPECINFO. Firstly, the user can input the proposed structure and calculate the carbon spectrum. By comparison of the calculated shifts and the real line positions one can immediately see whether the structure is reasonable. Alternatively, the user can input the chemical shifts of some or all of the observed lines and the computer can find

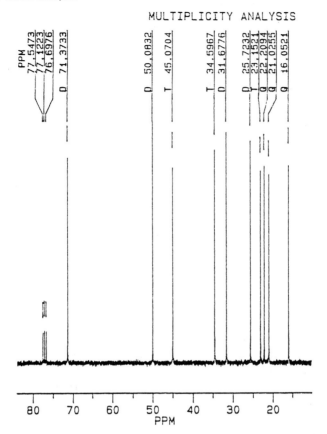

Fig. 11. ^{13}C NMR spectrum obtained during an automatic run on the spectrometer

and display all structures which match these lines. This analysis immediately gives an indication of the structure of the molecule under investigation. Additional information, e.g. any coupling constants or relaxation times which might be measured, can also be included in the search. If an unambiguous result is not obtained, other spectroscopic information (e.g. from mass spectrometry or infrared spectra) can be incorporated.

Figure 12 presents the result of a carbon spectrum calculation for the structure of menthol (**1**). The calculated chemical shifts closely match those observed in Figure 11. Figure 13 is an example of the input for a line search (taken from the experimental spectrum in Figure 11) and the corresponding computer output of structures which satisfactorily match the experimental data. The structures produced are very similar and the chemist can be confident that this structure is correct.

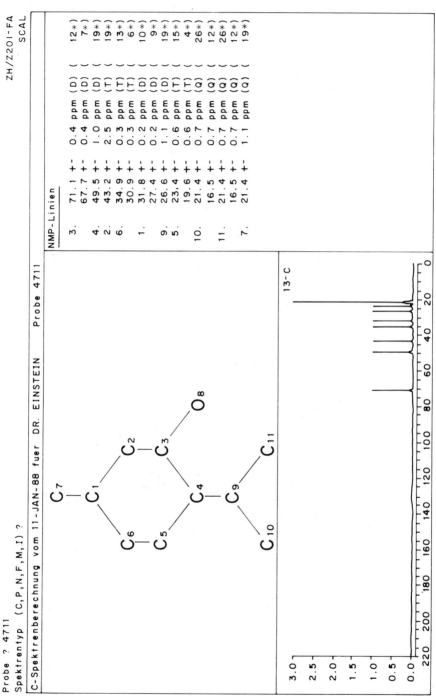

Fig. 12. Calculated ^{13}C NMR spectrum of compound 1

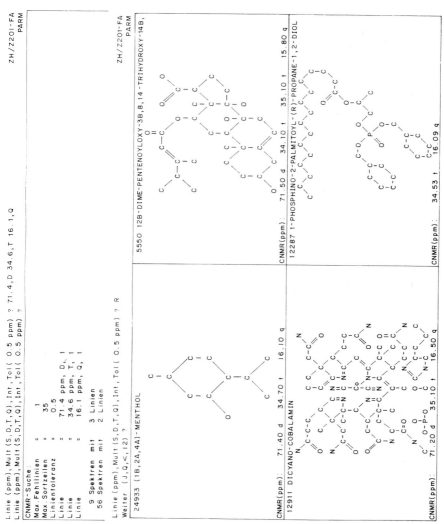

Fig. 13. Input for a line search with SPECINFO and the results

In our laboratory most of the routine carbon NMR spectra are interpreted with the aid of SPECINFO. A copy of the computer output is sent to the chemist, together with the spectrum. If the structure is new and all lines have been assigned unambiguously, the information is entered into the database. The database of SPECINFO is growing continuously and consequently the quality of the results is improving. Currently there are 86 000 different structures with carbon spectra stored with the SPECINFO database.

Carbon spectra of organic compounds normally show a number of lines. Information about the structure under investigation is extracted from the sum of the spectral information given by *all* of the observed lines. The situation is more difficult for spectra which show only one or two lines. These resonances typically span a wide chemical shift range. The user faces this problem in nitrogen or phosphorus NMR or in many other X-nucleus spectra. Because of the great shift range, the resonances of these nuclei are very sensitive to structural changes and the spectra contain a lot of information although they have only a small number of lines. Databases for heteronuclei assist in the interpretation of their spectra and can be used in the same way as the carbon database. In our laboratory about 2000 structures with phosphorus spectra and 1000 structures with nitrogen spectra are stored and used for spectral interpretation.

5   INTEGRATED AUTOMATIC SPECTRAL ACQUISITION AND INTERPRETATION: A PREVIEW OF THE NMR LABORATORY OF THE FUTURE

As shown in the preceding sections, sample preparation, acquisition and plotting of NMR spectra are now fully automated and powerful computer programs exist to assist in spectrum interpretation. Two main problems remain to be solved: firstly, the interpretation programs must be linked directly to the spectrometer, and secondly, the results of different spectroscopic methods must be combined and used in a structure generator to suggest novel structures.

The hardware to connect a spectrometer via Ethernet to a mainframe computer exists and is offered nowadays by all NMR manufactures. In the BASF central laboratory a Bruker AM360 spectrometer is connected to a MicroVAX II via Ethernet. Work is under way to develop programs for automatic input of spectral information to the SPECINFO program.

Figure 14 shows how data handling will appear in the near future. The chemist will send a request for NMR spectra by electronic mail to a mainframe computer together with a suggested structure and information about the sample. The sample still has to be delivered to the NMR laboratory by humans

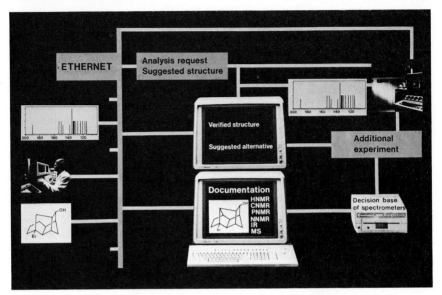

Fig. 14. Data flow in the laboratory of the future

and automation of this step will not occur in the immediate future. After the sample has been prepared (by a robot), it is transferred to the spectrometer and a routine measurement is performed (unattended). The spectroscopic data are then sent from the spectrometer to an interpretation program on a mainframe computer. Calculated chemical shifts are compared with the observed values and the results of a line search are compared with the proposed structure. If there are no ambiguities, the results including the spectra are copied by electronic mail to the electronic notebook of the chemist. In case the data do not allow the problem to be solved, a new experiment is chosen from a list of pulse sequences and parameters such as average relaxation times or coupling constants for the suggested structure can be taken automatically from the library and despatched to the spectrometer. The new experiment is initiated and the procedure is repeated. After a final result has been obtained (or the system decides that it cannot solve the problem), the next anlaysis is started. Some of the components, which are needed to embark on the next step of automation, exist already but work is still in progress.

As described before, NMR is nowadays automated to a high level. More automation will follow and this will allow the preparation, measurement, interpretation and documentation of even more samples with a given number of spectrometers and in a given period of time. Because of the higher throughput, with the same number of operators, the cost per analysis will continually decrease.

*"That does it! We're
replacing you with people."*

Fig. 15. Over-automated laboratory and incorrect use of computers

However, human operators will never be replaced. The automation and expert systems are developed to assist the expert to do faster and still better work and not to replace him. Otherwise, we might easily end up with a situation similar to that shown in Figure 15!

REFERENCES

1. W. Bremser, *Chem.-Ztg.*, **104**, 53 (1980).
2. S. Berger, *Nachr. Chem. Tech. Lab.*, **35**, 834 (1987).
3. W. Bremser and R. Neudert, *Eur. Spectrosc. News.*, **75**, 10 (1987).
4. (a) D. T. Pegg, M. R. Bendall and D. M. Doddrell, *J. Magn. Reson.*, **44**, 238 (1981); (b) M. R. Bendall and D. T. Pegg, *J. Magn. Reson.*, **53**, 144 (1983).
5. A. Bax and G. A. Morris, *J. Magn. Reson.*, **42**, 501 (1981).
6. W. P. Aue, E. Bartholdi and R. R. Ernst, *J. Chem. Phys.*, **64**, 2229 (1976).

Index